SOTHIS PRESS

Sothis Press aims to completely reimagine the public domain book with an emphasis on universal access to captivatingly designed print editions. We believe that technology will enable every single book ever published to be perpetually accessible in beautiful print form. Sothis is beauty in ubiquity.

www.sothis.press
1 Waterhouse Square, London, ECIN 2ST

Drilling for Placer Gold

Walter H. Gardner

SOTHIS PRESS

ISBN 978-1-0368-0809-9

This edition is published by
Sothis Press © 2023. The company
is registered in England and Wales.
The Company Number is 14900389.

5

DRILLING FOR PLACER GOLD

HARRON, RICKARD & McCONE

SOLE AGENTS

SAN FRANCISCO LOS ANGELES

Work continues in spite of snow

DRILLING FOR PLACER GOLD

KEYSTONE

KEYSTONE DRILLER COMPANY
BEAVER FALLS, PENNA.

A deep hole in the High Sierras, California

CONTENTS

CONTENTS—Continued

PART I

PLACER GOLD TESTING

PROSPECTING FOR PLACER GOLD WITH THE KEYSTONE DRILL

COMPILED BY

WALTER H. GARDNER

And after being critically reviewed by the original designer of this process and compared with the findings of a wide circle of authentic and practical engineers who have used the process, it is

PUBLISHED BY

KEYSTONE DRILLER COMPANY

Beaver Falls, Pennsylvania

DEDICATION

This Manual of Information for the Placer Engineer is dedicated to the hardy men who have carried the Keystone to the edges of the world; who have cleared a path for it with machete through the matted jungles; who have forgotten frost-bitten fingers in Alaskan Tundras to finish "one more hole"; who have lived and toiled in far lands and outlying camps that they might reveal new sources of treasure and open new fields of human endeavor.

PREFACE

"The churn drill is, however, the best device known for prospecting ground having the necessary conditions for gold dredging."

Engineering and Mining Journal

The Keystone Drill, as universally used, is not an instrument of precision. Gold particles are not distributed with mathematical symmetry. One drill hole to an acre in ground that is 50 feet in depth only yields for examination 1/164,000 part of the whole! Ordinary common sense and care on the part of the operator and panner are sufficient to insure acceptable field work. Meticulous precision or elaborate core measurements are generally absurd.

For after the depths have been recorded and the gold weighed, there comes the calculation of the values of precious metal in great blocks of gravel. There is no fixed formula. The experienced Engineer, to whom such work should be entrusted, will compensate for high variations, for loose and swelling ground, for sand and clay, for rusty gold—and his calculations will of necessity contain approximations that eclipse minor errors of the field and nullify minute measurements.

Indeed, field work with the Keystone Drill need not be conducted by men of profound skill. Just so the work is done in a consistent manner under the occasional eye of a competent engineer; just so the results are interpreted in the light of experience—then will the final figures carry the full weight of authority.

There is no mystery about field work—and should there be, the following pages will clarify the mode of procedure.

But there can be no standard of practice in interpreting the drill returns,—this book can here only serve as a manual of suggestions and reminders to the experienced engineer.

And there is herein contained a review of a generation of Keystone use in the light of the actual recovery from subsequent mining operations. Hitherto unpublished tables are offered for the files of all interested in the exploration for placer gold.

May you find this little book worthy of preservation!

THE AUTHOR.

CHAPTER I

THE ADVANTAGES OF THE KEYSTONE DRILL

KEYSTONE DRILLS HAVE CONFIDENCE OF ENGINEERS

To "Keystone" a gold dredging field is to "prospect" it! So closely has the Keystone Drill identified itself with the examination and calibration of auriferous gravels that its very name has grown into the jargon of the Engineer as a synonym of thorough and conscientious exploration. For the past twenty-five years have seen extensive areas of river gravels accepted for exploitation or cast into the discard on the strength of no other information than that revealed by the Keystone Drill in competent hands. With a serene confidence in its reliability, men have invested hundreds of thousands of dollars in the purchase of land and dredge machinery—or accepted an unfavorable verdict without a question.

WIDE RANGE OF USE

The Keystone Drill was first used for the determination of gold values in a placer deposit in Idaho, in the spring of 1898. Since that time many hundreds of drills have been shipped to all parts of the world and used in the testing of bench and bottom land; of old channels; of the bars of live streams; of lake and river bed. It has clattered in the depths of the Columbian jungles where it paved the way for the Nechi and Pato dredges; it has probed the frozen gravels

The Canadian Klondyke Mining Company used KEYSTONE DRILLS in their extensive exploration work

of Alaska and Siberia; it has charted the wide areas of the Oroville, Yuba and Natoma districts of California; it has explored innumerable of the lesser streams of Oregon and Montana, of the Philippines and the Malay States. It has sought for gold; for platinum; for diamonds and for tin. It has been transported under its own power; by horses; on the backs of mules; on snow-sleds; in crude dug-outs—even on the backs of human carriers. Sometimes a half dozen machines have been purchased and used to hasten the work on a promising property—there were at one time more than 40 Keystone Drills operating simultaneously on the property of the Lenskoi Mines. It has been used to determine the presence or absence of precious metals; it has been used to carefully appraise the contents of whole tracts; it has located limits of dredging possibilities both as to values, depths and bedrock reefs. It has proved that its results are the best index to dredging possibilities. And it has established a reputation for sturdiness and reliability that few machines enjoy.

KEYSTONE RESULTS GIVEN PREFERENCE

The Engineer who faces the task of correctly determining the recoverable gold content of a placer field will ordinarily have several alternatives of procedure. He may put down shafts. Open cuts or exposed walls may give access to the various strata. He may depend on a small and flimsy "hand-drill." Or he may elect to do a thorough and accurate exploration with a Keystone Drill. We briefly review the various shortcomings and advantages of these methods:

PROSPECTING WITH SHAFTS

Sub-surface waters usually prevent extending a shaft to bedrock. Uneven and improperly completed shafts do not yield reliable data. The same inaccuracy attends the attempt to sink a shaft below water level by the use of pumps or other appliances. The inflow of water will carry values. To be sure, if water did not interfere, a property might be thoroughly prospected by shafts and by shafts alone—but at an unnecessarily high cost. And while the final data would of course be accurate, it would, despite the greater size of the samples, be little more authoritative than the Keystone evaluations. So the Keystone Drill holes, miniature shafts as they are, are today recognized as preferable from the standpoint of the time and money that they save and as accurate to nearly the same degree. But the careful Engineer will usually check the operations of his Keystone with one or more shafts—to bedrock if possible and at least to water-level. He sinks the shaft around one of the early drill-holes. It gives him a better visualization of the strata that the drill has already indicated—it yields sufficient gold to obtain a test as to the nature and purity of the particles—it supplies some data for the decision as to the proper "constant" to be employed in his later calculations. And it also more surely indicates the exact level of the sub-surface water; it reveals something of the configuration of the bedrock; it more clearly tells of clay streaks and buried boulders. But the experienced Engineer will rarely waste time and energy on more than one such check shaft.

OPEN CUTS

A thorough exploration of a placer field will often discover open cuts, old shafts, exposed banks and other opportunities to cross-cut the gravels to some depth. But such exploration can seldom be more than casual and is either merely supplementary or in advance of the determination to drill thoroughly. For the random nature of such opportunities does not permit the systematic operations that alone carry the assurance of accuracy.

SMALL HAND-DRILLS

Standard equipment for use with the Keystone Drill is Extra Heavy Pipe 6 inches in diameter. It is generally held that a smaller sample cannot be error-free. Two and a quarter times as much material is recovered for examination from the Keystone Drill holes as from a 4-inch pipe. While there have been cases where the exigencies of transportation seemed to encourage the lighter machine, and while it may be permissable for preliminary work, it is of record that few dredges have been purchased without adequate Keystone results as the basis for the confidence of the investors. Indeed, it has become almost an axiom that capital will not be attracted to a placer field unless it has been thoroughly "Keystoned"! Nor can the "hand-drill" cope with conditions of hard ground. Dredge operators say that they can work any gravel that can be broken by a man with a pick—and the operations at Natoma, California, certainly prove the truth of this contention. But it requires a sturdy machine to probe such compact gravels—heavy enough to scorn the boulders that are inevitably encountered. The Keystone Drill is of as light a design as may be confidently entrusted to test with truth and accuracy the average deposit of auriferous or value bearing gravels.

In clover! Scene at Lincoln, Montana

KEYSTONE DRILL FITS NEEDS OF ENGINEER

The Engineer who wishes his work to carry the full flavor of accuracy or the promoter or property owner who wishes to collect data that will command the respect and belief of others will not be satisfied with any but Keystone Drill returns. Nor is the Keystone so heavy as to be difficult to transport or to operate. Quite the contrary. The Keystone No. 1, with the boiler sectionalized, may be transported by pack-train. The Keystone No. 3, the standard model for placer prospecting, is designed for speedy moving and "setting up"; for the rapid drilling of holes that are but rarely more than 60 feet in depth; for hard work far afield from machine-shop and organized repair facilities—in short, for the peculiar and arduous work of the pioneer *Placer Engineer!*

INCIDENT OF KEYSTONE USEFULNESS

There is on record an example of the untrustworthiness of shafts when they do not go to bedrock. One property was once prospected to the depth of about ten feet when water was encountered. These shafts revealed promising values which had only to be maintained to a reasonable depth to make an attractive dredging proposition. Indeed, so gratifying were the returns that a company was formed and stock sold. But cautious capital wisely demanded that the ground be Keystoned and it was discovered that the lower gravels were absolutely barren! It seemed that an old and filled-up lake bed, or settling basin, had been covered in recent times with an all too thin layer of auriferous gravel. Here the Keystone saved a very considerable sum of money for some one! Most Engineers prefer, if conditions at all permit, to rely on a Keystone even for preliminary work. It is disconcerting to arrive at a property after a long trip with nothing more adequate than a "hand-drill" for equipment and find ground that is too hard for it to attack and containing too much water for shaft sinking!

USE OF KEYSTONE WITH OPERATING DREDGES

There is yet another angle to the situation. In the event that preliminary work prompts a complete campaign of evaluation, the Keystone is absolutely necessary. And when the dredge is finally completed and at work, the Keystone has not outlived its usefulness. Modern practice suggests the continued use of the drill in front of the dredge, blasting hard ground, determining the boundaries of barren areas, charting the "pay-streak", seeking out reefs of hard bedrock, tracing the sub-surface bench line, even forecasting the current dredge returns. A large dredging company now follow this practice, for, as they say, "There is nothing more foolish or expensive than to prospect with a dredge!"

SUMMARY OF KEYSTONE ADVANTAGES

Here, then, are many reasons why a property under examination for its gold content should be Keystoned at as early a date as may be possible—

1. Keystone Drills are not too heavy to transport to properties that are at all accessible.
2. Keystone Drills are able to cope with severe conditions of deep bedrock, boulders and hard or frozen gravels.
3. Keystone Drills achieve very nearly the accuracy of shaft work at the cost of less money and time.

4. Keystone Drill returns command confidence and respect. Their use adds weight to any report on a placer property.
5. Keystone Drills are practically certain to be specified on the final examination—it is a matter of economy and efficiency to use them even for preliminary work.
6. Keystone Drills do not outlive their usefulness with the passing of the period of exploration.

CHAPTER II

PROPER DRILLING METHODS

THE CREW

Efficient operation of the Keystone Drill will in most cases require a crew of three men—the drill-operator, the panner and a fireman. Usually a man with a team is necessary to bring water and fuel as well as to aid in moving. Sometimes the Engineer in charge will prefer to perform most of the duties of a panner; sometimes, with oil-burning equipment, one man may be eliminated.

THE OPERATOR

The man who actually runs the drill should be experienced. On him largely depends the accuracy and certainly the expedition of the work. He should be capable of the repair, replacement and bit-sharpening which must all be sometimes conducted under the difficulties of limited conveniences afar from shop and machine equipment. The Engineer will largely be occupied with exploration, surveying, securing supplies, correspondence and manifold duties—his drill-operator should be a dependable lieutenant and a resourceful mechanic.

THE PANNER

The panner, also, should be an experienced, discreet and trustworthy man. His duties are to keep the records, check the core, collect the recovered gravels and slimes, clean up the sluice-box and make a full report to his superior on all drilling operations. He is often asked to supplement the operator's pipe and rope measurements with an independent check of his own. But the ideal man will do more than that —he will turn a willing hand to whatever chore presents itself, for he will have unoccupied moments. It is of importance that he be provided with shade in hot weather and shelter and warmth in cold; otherwise his work can not be kept free from variables incidental to cold and discomfort and can not be prosecuted with full accuracy—it being unavoidable that his hands be in water much of the time.

THE FIREMAN AND THE WATER-BUCK

The work of keeping up steam and driving the team may be entrusted to such labor as can be secured in the field of operations. Ordinarily such men need not be skilled to any degree—but the very nature of prospecting work suggests that there be some assurance of the loyalty and discretion of every employee. The fireman should be required to start his work on each shift in such season that a full head of steam will be available upon the arrival of the rest of the crew.

FOUR-MAN CREW

In the extensive operations of one large company, the employment of a fourth man has been found a justifiable expense. With such a crew it becomes the duty of the fireman to care for the fuel, water and lubrication; to act as mechanical inspector, keeping all bolts tight; and to keep a full head of steam. The fourth man, or "helper", acts as assistant to both panner and operator; cleans casing threads; prepares roads in advance of moving; runs needful errands, and in general serves to keep lost time at a minimum.

In a Far Corner of Manchuria

SELECTION OF MACHINE

For these many years the Keystone No. 3 Traction has been the favored Model for placer prospecting. It is amply powerful for all depths and all gravels likely to be encountered. The friction-hoist is to be preferred. The No. 3 Non-Traction may be used, of course, but the traction machine soon earns back in saved time its somewhat greater cost. Where transportation into regions difficult of access must be provided, the lighter Keystone No. 1, with its boiler sectionalized, may be carried in by pack-train. On those rare occasions when the holes run beyond 300 feet in depth, the Keystone No. 5 is specified.

FUELS

The standard fire-box of the Keystone Drill will take wood of about 18 inches in length by 6 to 8 inches in thickness. Seasoned wood is the best, of course, but a clever fireman can keep up steam with damp and green fuel that would appear almost useless. It is often possible, and always desirable, that supplies of wood fuel be prepared in advance. Coal may be used interchangeably with wood, the Keystone grate being suited to either, but a supply of extra grates is advisable if coal is to be burned. The machine can easily be equipped with an oil-burner where proper fuel is securable—and when so fitted the costs of operation are materially lowered. In many parts of Alaska, Keystone Drills fitted with gasoline engines serve best. The drill will use a little less than half a cord of wood to a ten-hour shift if the wood be of average worth, and the operations require four to six barrels of water.

EXTRA EQUIPMENT

There are listed hereafter not only the standard equipment for placer testing, but also a selected list of extras and spares, with the accessories, that intimate acquaintance with field conditions prescibe as proper and necessary adjuncts to speedy and efficient work. Let it be borne in mind that the search for gold is carried on in the far places of the world, and that, except in those few instances where operations are amid the conveniences of civilization, placer prospecting parties must rely largely on their own resources. Indeed, it is the ruggedness of the Keystone Drill and the simplicity of its upkeep that have raised it so high in favor! Again, speed in carrying out the work is invariably to be desired; there is always an overhead expense in maintaining the crew that counts heavily in the event of lost time. The simple precaution of carrying a complete outfit and ample replacement parts will insure against excessive expense and irksome delay. The drill-bits must be sharpened in the field—that must be provided for. Sand pump valve packings wear out—even an extra valve assembly should be carried as precaution against embarrassment due to breakage. Continued use eventually "fatigues" the bolts for the driving clamps. Connecting-rod brasses require replacement—there are many possible mischances of minor nature incidental to the strenuous work of drilling into impacted gravels. It will be found that the Keystone list of extras as appended cannot be safely curtailed if there is to be provision against ordinary hazards; and, indeed, it may well be increased if the expedition is likely to work for a long time far from the beaten track. A complete set of carpenter and blacksmith tools will then be found a worthy addition.

PACKING AND SHIPPING

For domestic use the Keystone Drill will generally be shipped completely assembled. For use abroad or in distant fields it may be partially dis-assembled and crated. The new machine, ordered from the branch of the Keystone Driller Company, will be carefully inspected to insure immediate readiness for operation. A machine once used, that is to be transported to a new field of work, should be carefully and thoroughly looked over by the Engineer so that possible delay or confusion at time of delivery may be obviated. The machinery and tools should be checked as complete and in perfect working order. Packages and crates should be numbered and listed to permit of ready identification. All should be plentifully greased to prevent oxidation in transit. The rope should be protected from moisture. Steam engine oil and greases should be included in the shipment if there is any uncertainty as to the availability of these in the field. Four 3″ by 12″ planks are often shipped with the drill, and prove useful in unloading, moving and blocking. The sluice-box can be made in the field if a piece of 30″ by 60″ Galvanized Iron, 20 Gauge, be rolled up and included. The rocker had best be fabricated on the ground from the best available soft pine—for it is almost sure to be dried out or broken in transportation. Wise forethought will yield large dividends in saved time and lowered cost of work!

DRIVE PIPE—CARE AND KIND

The pipe recommended for placer testing is the Extra Heavy Drive Pipe (28 pounds to the foot), for the lengths are driven and pulled and used over again many times. A generous supply is most desirable—on far expeditions three

A KEYSTONE and its quaint Chinese Crew

times as much as would be required for one average hole may well be purchased and transported. Short lengths are advised for greater ease in handling—an accepted length is about six feet for each piece, though the first joint may well be seven or eight feet. The threading must be accurate and true; that there be no waste of time in screwing and unscrewing; that the joints will butt in the middle of the coupling, thus avoiding danger of thread-stripping, and so that all pieces will be interchangeable. Threaded ends should at all times be protected by couplings or short sleeves. Pipe ordered from the Keystone Driller Company is threaded to the "Keystone Standard Cut."

DRIVE PIPE—LONGER LENGTHS

However, if the holes are to be more than 60 feet in depth, longer lengths are advisable. To permit their use, provided that the ground is dry, the hole is started in a pit about 6 feet in depth and 5 feet by 7 feet in width and length. This pit is cribbed lightly at the top to prevent the jar of the operation of the machine from breaking it down; it is covered with substantial planks and the weight of the drill is supported on wide and heavy boards. With such a pit 10-foot lengths may be comfortably used. In wet ground where timber is plentiful the drill is often cribbed up so as to be 4 to 6 feet above the surface and the same lengths may be employed. There are then fewer joints to provide the possible cause of a "crooked hole" when the gravel is compact, and less time is taken up in making pipe connections.

DRIVE PIPE—INSPECTION

In the field it is customary to use certain pieces of pipe in a regular sequence, numbering them for identification. It be taken to examine the material removed for possible values, although the hole will usually be started in barren silt or loam. The stem is then lowered into the pipe and the driving-blocks bolted on. The attendant, mounted on one of the cross-pieces of the derrick, will steady and balance the otherwise unsupported stem, while the panner will hold the pipe until driven to a depth where it can support itself. Ordinarily, in surface soil the first drive may be for two feet, when the depth and core should be read off and recorded. A little water may then be added (not too much, on account of the possibility of its washing down through the loose soil) and a pumping made *without* drilling. In hard compact gravel it is sometimes necessary to drill first—but this is a procedure to be avoided if it is at all possible.

USE OF WATER

The contents of the pipe should be kept thoroughly mixed with water as the work progresses. In dry ground no more should be added than is necessary—perhaps three or four feet of it in the pipe. But in wet ground the water-level should be kept *above* the water-level outside the casing. This higher head has been found to be efficacious in largely preventing the inrush of too great a core and the intrusion of extraneous values. The Engineer will sometimes find this to be a troublesome point, but one that must be insisted upon. The operator should continually have before him the fact that he is not there to make progress, *but to gain a correct sample!* Care must also be taken to flood the drill-stem with a bucket of clean water as it comes up after the drilling and before the pumping, and to give the sand-pump a final flush on the conclusion of each pumping. The boiler water should be kept in a separate barrel—where no dirty buckets are plunged into it. The water used in the drill hole should be absolutely free from oil or grease of any kind.

DRILLING—TIGHT GROUND

The pipe will have been carefully measured and marked at intervals of 6 inches with a water-proof crayon. The assembled drilling tools will also have been measured and marked so that at any instant the operator will know the exact relation of the edge of the bit, at the limit of its downward stroke, to the bottom of the cutting-shoe. Drilling will proceed until the bit is from 1 to 4 inches above the lower end of the cutting-shoe; this thickness of impacted gravel prevents the intrusion of too much material from without the path of the pipe. This thickness of gravel is alluded to as the "core". The drill-stem should be turned slightly by the hand of the operator to keep the blade from striking in exactly the same place and packing the contents of the pipe or becoming stuck. If a boulder is encountered it may be broken up by drilling *ahead* of the pipe—proper notation being made in the log. Drilling below the pipe is always a bad and a dangerous practice in placer testing, except that in heavy ground or coarse gravel it may be necessary to drill a few inches ahead after pumping to clear the pipe of a plugged core. Before resorting to this expedient, it should be ascertained that the water-level in the pipe is above the water-plane in the ground. Fortunately, in such heavy, tight gravel there is ordinarily found a stiff clay which serves to protect the walls of the hole below the drive-shoe.

be taken to examine the material removed for possible values, although the hole will usually be started in barren silt or loam. The stem is then lowered into the pipe and the driving-blocks bolted on. The attendant, mounted on one of the cross-pieces of the derrick, will steady and balance the otherwise unsupported stem, while the panner will hold the pipe until driven to a depth where it can support itself. Ordinarily, in surface soil the first drive may be for two feet, when the depth and core should be read off and recorded. A little water may then be added (not too much, on account of the possibility of its washing down through the loose soil) and a pumping made *without* drilling. In hard compact gravel it is sometimes necessary to drill first—but this is a procedure to be avoided if it is at all possible.

USE OF WATER

The contents of the pipe should be kept thoroughly mixed with water as the work progresses. In dry ground no more should be added than is necessary—perhaps three or four feet of it in the pipe. But in wet ground the water-level should be kept *above* the water-level outside the casing. This higher head has been found to be efficacious in largely preventing the inrush of too great a core and the intrusion of extraneous values. The Engineer will sometimes find this to be a troublesome point, but one that must be insisted upon. The operator should continually have before him the fact that he is not there to make progress, *but to gain a correct sample!* Care must also be taken to flood the drill-stem with a bucket of clean water as it comes up after the drilling and before the pumping, and to give the sand-pump a final flush on the conclusion of each pumping. The boiler water should be kept in a separate barrel—where no dirty buckets are plunged into it. The water used in the drill hole should be absolutely free from oil or grease of any kind.

DRILLING—TIGHT GROUND

The pipe will have been carefully measured and marked at intervals of 6 inches with a water-proof crayon. The assembled drilling tools will also have been measured and marked so that at any instant the operator will know the exact relation of the edge of the bit, at the limit of its downward stroke, to the bottom of the cutting-shoe. Drilling will proceed until the bit is from 1 to 4 inches above the lower end of the cutting-shoe; this thickness of impacted gravel prevents the intrusion of too much material from without the path of the pipe. This thickness of gravel is alluded to as the "core". The drill-stem should be turned slightly by the hand of the operator to keep the blade from striking in exactly the same place and packing the contents of the pipe or becoming stuck. If a boulder is encountered it may be broken up by drilling *ahead* of the pipe—proper notation being made in the log. Drilling below the pipe is always a bad and a dangerous practice in placer testing, except that in heavy ground or coarse gravel it may be necessary to drill a few inches ahead after pumping to clear the pipe of a plugged core. Before resorting to this expedient, it should be ascertained that the water-level in the pipe is above the water-plane in the ground. Fortunately, in such heavy, tight gravel there is ordinarily found a stiff clay which serves to protect the walls of the hole below the drive-shoe.

Drilling in China behind the protection of a fence and barricade

DRILLING—LOOSE GROUND

If there is loose and running ground there may be but little drilling necessary, and the utmost skill and experience of the operator will be called into play to prevent the pumping of too much material. In fact, there will frequently occur a stratum where the amount of silt and sand pumped out of the hole will be greater than the theoretical displacement of the drive-shoe—a careful recording of volumes will then give the Engineer data for the approximations and allowances that must be made. Fortunately, such loose material seldom contains heavy enrichment. If too thick a core is maintained, there is a possibility of its becoming a plug in the pipe and crowding other material aside as it is driven. The whole science of accurate drilling consists in pumping as nearly as possible the entire material in the path of the pipe—*and no more!* No two pieces of ground will react to drill operations in just the same way; indeed, data have been collected which show that no two skilled operators will secure exactly parallel results in the same hole! Here, as elsewhere, crops up the personal factor which the Engineer will study and reflect in his final calculations. In most work, however, variations will be so slight as to be ignored, or will largely compensate.

USE OF JARS

After some depth has been reached, or where boulders threaten to "stick" the drill bit, there may be occasion to put on the jars—not only to afford the impact necessary to loosen the bit by the jerk they can exert, but also for their added weight. Generally, however, they are quite unnecessary.

PUMPING

When the operator has broken up and loosened the material nearly to the bottom of the pipe he will throw the crank wheel out of gear and the cable reel into gear, thereby hoisting the tools clear of the pipe, and will swing the stem out of the way—usually throwing it over and behind the cross-arm. One pail of water should be dashed over the stem as it rises; more water should be added before pumping, as previously described. The Vacuum Sand Pump is then dropped into the hole, and rapidly raised and lowered two or three times to suck in the material. This pump contains a plunger which travels through its whole length. The rapid dropping of the pump forces this down and when the sand reel is thrown into gear, this "sucker" is drawn up so rapidly as to produce a vacuum in the lower part of the pump, thus opening the valve and drawing in the loosened contents of the pipe. The secret of successful pumping lies in imparting a properly rapid motion to this plunger and the power and control of the Keystone Drill enable the operator to recover the gravel, slimes, sand and mineral enrichment with a remarkable thoroughness. Two pumpings are usual in ordinary work where progress is made a foot at a time. Often time can be saved by pumping immediately after driving, drilling only when there remains more than 2 or 3 inches of core. The two pumpings, the one before and the one after drilling, are preferably caught in the same pan and concentrated in one operation. When extraordinary care is not required the material may be poured directly into the pan which will have been placed across the frame of the sluice-box, the slimes being allowed to accumulate in the box for the final "clean-up" on completion of the hole—the panner making proper record of the values therein contained. This sluice-box, a wooden frame with a half-round trough of galvanized iron, is mounted on wooden supports directly in front of the drill and extending forward away from the pipe. The attendant will carry the valve end of the pump out from the pipe as the operator easily slackens the cable (not enough but that the line still carries most of the weight of the pump) until he can lay it over the box; then he will raise the valve end and dump its contents into the pan. The pump should then be washed clean, both inside and out, with clear water and allowed to rest on top of the sluice-box (so that sand-line and drilling cable do not touch) ready for the next pumping.

CHECKING VOLUME OF MATERIAL PUMPED

If variable cores and unusual conditions suggest a greater accuracy in the observation of the relation of the recovered material to theoretical displacements, the entire pumping may be at once emptied into the sluice-box. At its lower end will be placed a large tub of water containing a pail into which the material falls. The contents of this pail will be measured before panning—the tub will gradually collect the slimes for the final accounting. The pail had better be calibrated and a chart prepared showing the cubic feet of material for each inch which the surface lies below the top of the pail. There is here given (Plate "E") an actual record of such data as presented in the report of a well-known Engineer.

DRIVING

After the pumping, the drill-stem will be swung back into the hole and carefully lowered to rest on the surface of the

Bit sharpening at Featherville, Idaho

material within the pipe. A measurement should then be taken to ascertain the amount of material above the drive-shoe and record made thereof as "core after pumping." The drive-blocks are then bolted on and the pipe driven the proper distance, usually a foot, more in loose and easy ground where no values are anticipated; less in very hard and refractory gravel or resistant sand. It has already been noted that there are occasions when lack of progress indicates that one may reluctantly drill ahead to facilitate driving, but the Keystone Drill has ample power to cope with exceptionally hard formations and the logs of Keystone drill-holes are singularly free from record of this expedient.

FINISHING HOLE

Most holes are continued to the hard rock, shale, decomposed granite, tufaceous lava, or barren stratum that marks the limit of recoverable values. The Engineer will usually prefer to be present at the conclusion of a hole, not only on account of the greater concentration of material usually encountered on bedrock, but to observe what he may of the character of this formation. The pipe should be driven to a depth that makes certain of bedrock and of the total absence of values—but not to a depth into the bedrock that will make pulling a slow and tedious process. The full purpose of the test hole having been supposedly now accomplished, the operator may, at will, drill a few feet into the bedrock. This will prove whether his supposed bedrock is or is not a large boulder.

PULLING PIPE

The long stroke of the Keystone Drill, together with its responsive control, facilitates the withdrawal of the heavy drive pipe for use over and over again. The stem is detached from the rope-socket and the "puller" screwed on tightly. The puller knocking-head is then attached to the top length of pipe and the machine started with the slack so controlled as to strike a forceful upward blow. If the pipe yields but slowly, the pulling ring and lifting jacks may be employed. The water-level in the hole is usually measured during this operation, although, if the hole "stands up", more nearly accurate information may be obtained by delaying this reading for a day or so. When the last joint is removed, the drill-stem is again fastened to the rope-socket, loaded into the bed of the machine, the jacks removed, and the outfit moved to the next hole.

PANNING

Under ordinary conditions the panner will concentrate the recovery of each pumping in his pan. After removing all but the minerals and black sand, he will estimate the weight of gold in milligrams and also count and record the colors. It is customary to record these as of the "First", "Second" or "Third" grade. A "Third" grade color is one that is large enough to be individually counted, yet below two milligrams in weight. A color between 2 and 7 milligrams is of the "Second" grade—above 7 of the "First" grade. This counting of colors and estimate of weight serves several purposes. It shows the occurrence of values according to the depths; it gives data for the Engineer if the recovery of a pumping is to be later discarded or discounted; it gives some idea of the worth of the ground if by accident the recovered values are lost or contaminated. The gold from the various pannings as the hole progresses will be accumulated in the same globule of mercury. This, preserved in a glass tube, will be kept on the person of the panner. On finishing the hole the tailings will all be re-run through the rocker, together with the slimes from the sluice-box and tub, and any values added to the mercury. This will then be handed to the Engineer, together with the log of the hole which it is also the panner's duty to keep. When the drilling is being prosecuted by a night-shift it has proved unwise to attempt panning by artificial light. Under such circumstances the pans are preserved for concentration on the following day, being marked by a wooden "paddle" on which the proper identification of the pumping may be pencilled. The panner is held responsible for the correct recording of the progress of the work, of the cores and for the recovery of every particle of metal.

TIN PROSPECTING

The procedure in evaluating tin placers is practically the same as for gold—except that the panner will screen his materials and grind the larger particles of tin to determine a sample and gain an assay that will declare as to the proportion of dross in the heavy tin-stones. Placer gold particles are ordinarily unadulterated with quartz—not so with tin "nuggets".

PLATINUM

Inasmuch as platinum is not picked up by the mercury, it is necessary to save for assay the heavier concentration of the pannings. In rare cases, the platinum will represent no small proportion of the values—in all placer fields it is a wise precaution to make a careful test of its presence in commercial quantities. For the ordinary gold-saving, tables of the modern dredge make a remarkable recovery of these heavy particles.

FIRE ASSAYS IMPROPER

Every Engineer has heard of instances where through ignorance the concentrate from *auriferous* gravels has been sent to the assay office for a fire test. This grave error is even today not uncommon. It should be remembered that the heavy quartz pebbles carry gold that no *washing* device can recover—and that while field panning and hand concentration methods may be crude, that they are of comparable efficiency to the gold-saving devices of the dredge and that more nearly accurate recovery from the sample will be dangerous and misleading. The "black sand" concentrate from the bottom of the pan, after the mercury has picked up all visible yellow particles of gold, will often yield a high fire assay—but the recovery of these values by washing or amalgamation is even more nearly impossible on a large scale than it was in the prospector's pan!

PRECAUTIONS TO BE OBSERVED

On leaving the work at the end of a shift, it is an excellent idea to throw into the hole a few handfuls of tailings that are known to be barren. This is pumped out and inspected on beginning the next day's work—an effectual safeguard against tampering. In fact, it is impossible to salt a Keystone Drill hole if even moderate watchfulness is employed. It is rumored that on one occasion particles of virgin gold were inserted into the fibres of the drill-rope, these gradually falling into the hole as the work progressed. The sight of a few flakes of bright gold on the top of the driving-head revealed the trick. It is also said that once an interested property owner loaded a cigarette with precious dust and flicked the ashes into the pan as concentration was being carried on. Salting is only possible by some such fantastic and wholly unlikely ingenuity. However, the careful Engineer will from time to time check the purity of the quicksilver which he uses.

FROZEN GROUND

A new technic of drilling in the frozen gravels of the Arctic has been developed. This work requires experience if it be done with accuracy; we can only touch on the method employed. When the pipe reaches frozen strata, drilling and pumping and panning proceed *below* it—a foot at a time. There will be some caving, and an irregular hole is pretty sure to result as thawing proceeds. After bedrock is reached, the tools are withdrawn and a measured quantity of water kept at a temperature just above freezing is poured into the hole. A careful record is kept of the volume necessary to raise a float one foot vertically—and thus is secured an equivalent measurement of the cubic feet of gravel actually removed in each foot of progress. With the new science of "cold-water thawing" and the renewed interest in Arctic gravels, this ingenious method of insuring reasonable accuracy in prospecting becomes particularly noteworthy—and here again is the power of the Keystone Drill of signal value!

UNUSUAL CONDITIONS

The vast differences encountered in probing the placers of the world cannot all be dwelt upon in this short treatise. Indeed, there will be unprecedented conditions now and then arising that will tax the resourcefulness of Engineer and operator alike. But the ruggedness and yet splendid flexibility of the Keystone Drill adapt it to the hardest tasks. It can be cribbed up high above marshy ground when long

Bit sharpening at Featherville, Idaho

pipe lengths are to be handled. It can be operated from a scow afloat in lake or river. It is a sturdy machine for hardy men in rough country!

Here might be introduced a paragraph on the Keystone as a Shaft-Sinker in frozen strata. The advantage and process are that:—A six-inch hole is first drilled through the frozen strata. Two or three returns of one-inch pipe connected to the boiler serves overnight to thaw the material out to shaft size. The drill-stem having been removed, the cable and cable reel are, without other change, used to elevate the mining bucket for the thawed material or the accumulated water. The machine and its accessories in fact make it a complete shaft-sinking outfit.

CHAPTER III

LAYING OUT THE GROUND AND ESTIMATING THE VALUES

SURVEY

The careful examination of a placer deposit requires a thorough survey and an adequate map—this being made with sufficient accuracy and to so large a size that distances may be scaled with precision. It is not always possible to secure such a survey in advance—the Engineer will often prosecute such work while supervising the beginning of drill work.

WHAT MAP SHOULD SHOW

On the map should appear property lines, roads, bedrock outcroppings, and physical limitations of dredging areas; and it should represent a complete picture of the property not only for purposes of prospecting, but for later use in buying parcels of land and for dredge operations. Perhaps such thorough work will not be completed until there is some assurance as to the desirability of actual exploitation, but the initial survey may quite as well be conducted so as to be made full use of if the property "proves up".

PLOTTING DRILL HOLES

Usually the first drillings will render a verdict as to the ultimate value and dredgability of the property. Accordingly, it is often desirable that they be well scattered so that they may roughly reveal the worth of the entire acreage. Yet, if the results are favorable, this first work should have been so done that it fits into the complete campaign of prospecting.

EXPLORATION WITH DRILL

In a channel deposit the Engineer will often stake out lines at right angles to the stream flow and mark proposed holes just as if he were assured of favorable results. Then he will start drilling by driving alternate holes on alternate lines—with the purpose of later completing the work when the value of the area has been reasonably established. In a "blanket" deposit he may chart his entire scheme of drilling —then actually drill here and there over the entire acreage until a complete exploration has been justified.

LAYING OUT HOLES IN STREAM DEPOSIT

In all ordinary deposits a Keystone Drill hole to every two to four acres will be adequate to yield complete data as to values. The plotting of proposed drill-holes will be made with this in mind. Several "channel" or stream deposits have been appraised by drilling holes at a distance of 150 feet on lines that are 1000 to 1500 feet apart. The location of holes on Plate "A" is typical.

"BLANKET" DEPOSITS

As is hereafter described, the marking of a wide acreage of auriferous gravels into equilateral triangles, placing drill-holes at their apices, is an excellent method of conducting an exploration. The final calculations are then quite simple and the chance of confusion and error is eliminated. Some Engineers prefer to plot the field in rectangles, drilling at the corners of each parcel. Still others divide the field into squares and drill at the center of each square. This brief treatise does not attempt to argue the merits of various methods but to fully present one method for the uniniti-

Everything running smoothly—even the Fireman smiles

ate—and record the several angles of consideration as memoranda for the more experienced.

TREATMENT OF AMALGAM

The Engineer will receive from the panner a small phial in which he has preserved the mercury containing the gold from the sample. This he will treat with dilute nitric acid (Sp. Gr. 1.42 plus an equal volume of distilled water) in a test tube over the flame of an alcohol lamp—wash well with boiling water to which a few drops of alcohol have been added—dry and anneal in a small annealing cup or porcelain crucible and weigh the clean dust on scales that are accurately sensitive to a milligram. Pocket scales are made that are sufficiently precise for this work. These operations should be carried out in a room free from dust or air currents and with clean reagents and containers. Then the Engineer, with the weight of the gold and the log of the hole before him, proceeds to the calculation of values in "cents per cubic yard"—the accepted basis for the evaluation of placer properties.

DETERMINING FINENESS OF GOLD

To properly translate his gold values he must appraise the fineness of the metal peculiar to the property in question. Many Engineers insist on sending a sample of the gold collected to the nearest available authority on quantitative analysis to determine with accuracy the proportion of gold and silver. Sometimes, if platinum be expected as occurring in the sands, as large as possible a sample of the panner's concentrate is also sent and a determination of that metal likewise requested.

VALUE OF MILLIGRAM OF GOLD

A customary approximation, when accurate data are lacking, is to call a milligram of gold worth .06 cents. This is based on gold that assays $18.66 worth of pure metal to the troy ounce of 480 grains—one grain being equal to 64.8 milligrams.

CALCULATION OF CUBIC CONTENTS OF DRILL-HOLE

A 1 foot hole yields .3068 cubic feet of material. If that material produces 1 milligram of gold then a cubic yard will produce 27/.3068 times as much—or 8.8 milligrams. But, on a basis of gold that is worth $18.66 per ounce; one milligram is worth .06 cents. Therefore a hole one foot in depth that yields one milligram of gold proves a value of 8.8 times .06—or 5.280 cents. Then, since this figure represents the yield of a hole of the unit depth of one foot and the unit recovery of one milligram we have only to divide it by the depth in feet and then multiply it by the number of milligrams recovered to learn the indicated values in cents per cubic yard for any hole! A table is included in this booklet which reduces this arithmetical calculation to one operation. (Plate "G"). It is based on the working out of the following formula:

$$\frac{27 \times .06}{.3068 \times \text{"D"}} \times \text{Mgms.} = \text{Value in Cents per Cubic Yard.}$$

.3068	Equals area of drive shoe in square feet.
27	Equals conversion factor to cubic yards.
.06	Equals value of gold in cents per milligram.
"D"	Equals recorded total depth of hole in feet.
Mgms.	Equals number of recovered milligrams.

In this appended table there have been substituted various depths for the factor "D"—there remains only to multiply the partial result opposite to the actual depth to bedrock by the number of milligrams to find the indicated value in cents per cubic yard.

COMPENSATION FOR EXCESSIVE CORES

Occasionally a hole will be reported that varies in a marked manner from normal. For instance, suppose that between the 20 and 25 foot marks there was trouble with a defective pipe joint or a plugged core that kept out the proper amount of material or an inrush of far too much sand and gravel. The panner's log would show an estimate of 18 milligrams of gold in this five feet of progress and extracted from either too little material or too much. Suppose that the amount of material actually recovered and the theoretical recovery do so fail to coincide, what then? Some Engineers prefer to throw out this portion of the hole; it is surely not necessary to discard the whole drilling by reason of this mishap, nor yet should such unreliable data be incorporated as part of the whole. In this hypothetical case the Engineer would subtract five feet from the recorded depth and 18 milligrams from the weight of the gold as finally wieghed; then calculate his values. Other Engineers will credit values to this five feet of hole according to the average of a like distance above and a like distance below the error. This seems a likely way to approximate the truth.

OTHER CONSTANTS IN COMMON USE

But the constant of .3068 (based on the calculated volume of a cylinder 1 foot high with a base 7½ inches in diameter) is only one of several that have been widely used in the past and are more or less favored today.

CONSTANT OF .3333

It was once a favorite practice to estimate values on the above basis and then to discount the completed returns on the appraisal of the property by 10 or 15 per cent as a measure of safety and conservatism. However praiseworthy this desire to err only on the side of safety, this procedure is hardly professional or sensible. At least one serious error has occurred where the values were so discounted *twice* through a misunderstanding. But this desire to lean to the safe side led to the adoption of an arbitrarily fixed constant of .3333. This was widely used by those who insisted on ultra-conservatism in the estimating of dredging values and also by those who maintained that the beveled edge of the Keystone drive-shoe as well as the entire method of operation allowed more material to enter the drill-pipe than strict theory would indicate. The use of the constant .3333 gave final values that were 8.7% less than when the .3068 factor was employed.

KEYSTONE CONSTANT OF .27

But from the very earliest day of the development of the technic of prospecting, there have been many Engineers who have favored the constant of .27 (one cubic yard of material to each 100 feet of pipe). This has been widely called the "Radford Factor" or the "Keystone Constant." Its use was first scouted as too likely to exaggerate values, but in the light of recent comparisons of estimated prospecting totals and the actual recovery of subsequent dredge returns, this "Keystone Constant" climbs back into secure favor. One large Company that has successfully operated many dredging properties in different parts of the world for a long time preferred the conservative constant of .3333—but it is noteworthy that their estimates today are usually worked up on the basis of the "Keystone Constant."

REASONS FOR THE KEYSTONE CONSTANT

The churn drill, functioning by impact, continually acts to drive the gravel downward before the pipe and to settle the heavier particles of gold and precious metal. If any metal ever does escape the action of the pump, it is more likely to be these same heavier particles. At the conclusion of the hole the last drilling may have driven a few particles of the metal down out of reach into the bedrock. The panner may always, despite the most scrupulous care, lose a tiny proportion of the values from the splash of the pumping or in the rocker. There is more likelihood of missing some of the values than of getting more than the due amount. There are many sound reasons to justify the practice of crediting the recovered gold to a volume of material somewhat less than the theoretical amount—and the "Keystone Constant" does just that.

NO ONE CONSTANT INVARIABLY PROPER

Few Engineers will use the same constant under all conditions. For there are many variables to consider in translating the work of a drill to a well considered prophecy of the values that the dredge will recover. The Engineer, in choosing his constant will weigh many things. The personality of his crew cannot be ignored. The time of the year in which the work is done will have an effect—a panner cannot work in freezing weather with the same accuracy as in the pleasant days of summer. Rich and streaky channels of

Pumping. A scene from work in Idaho

heavy gold concentration will usually drill higher than the expected recovery—a small gold content evenly distributed in loose ground will often show to the drill as less than what the dredge will later actually save. These matters will be discussed more at length in the final chapter of this article—"Chapter IV—*Reliability of Keystone Samplings*". Most Engineers today will use the "Keystone Constant" as the proper basis for their estimates.

COMBINING VALUE OF VARIOUS HOLES

When the Engineer has correctly evaluated the various holes he faces the problem of combining them to gain his tract values. It is customary to plot the holes on a large map, noting dredging limits as determined by surface contour, by exposed rimrock or reefs—and also marking property lines. If the values fade off into barren areas there should be marked the probable boundaries of dredging values—roughly estimated at the probable cost of the dredging operations. This may vary from 5 cents where a large dredge is to work under the most favorable conditions to 15 cents or even more where inaccessibility, expensive power or frozen ground offers difficulties. If the ground shelves to shallower depths the limiting depths should also be plotted. These vary for different dredges, but the average dredge can hardly dig its flotation in ground that is not at least 12 feet in depth. The yardage should be estimated from this completed plat with all possible accuracy. It is perhaps true that a little carelessness or a wrong conclusion at this point will affect the figure of total recoverable gold more than any other error that is likely to be made. It may be noted in passing that most Engineers add to their yardage calculations the amount of ground that the dredge buckets will handle in digging bedrock to the depth of one foot. This is a necessary procedure in actual dredging and one that dilutes the auriferous gravels with barren material.

PRINCIPLE OF EVALUATION

The only principle involved in calculating the value of a tract of ground lies in the necessity of giving proper "weight" to each drill-hole in that area—that is, in letting a drill-hole affect the final result in proportion to the area which it represents, or "governs." Merely to add the values of the holes and divide by the number of them is an elementary error.

CALCULATION WHEN HOLES ARE SPACED EQUIDISTANTLY

If the area of auriferous gravels is a "blanket" deposit and it has proved feasible to lay out the holes so that they are equi-distant, this calculation is a simple one indeed. The ground is then divided into equilateral triangles, and the depth of each area, and its values in cents per cubic yard is the average of the product of the depths and values of each hole. Thus, the deeper the hole, the greater its effect on the values of the area. The value of the tract is the average of the products of the volume of the triangles and their values. If the holes are not evenly spaced, to connect them by lines, to figure the area of the enclosed triangle and to allot to it the average of the products of the depths and values of the three holes at its apices is not far wrong. To be sure, there is a slender factor of error that increases as the "triangle" departs from the equilateral, but this is a compensating variation that may often be ignored.

OTHER METHODS

There are Engineers who favor other and more complicated calculations—but the principle to be followed is a simple matter of common sense that lends itself to translation in ordinary arithmetical terms. Discussion of the infinite number of special problems that the prospecting Engineer will meet can only bring one back to the starting point. It has been shown by actual test that rarely will two Engineers go about combining the results of two or more drill-holes in exactly the same way. But there is little chance to go far wrong. If there be a large area and many holes, the result will surely be within the limits of error of the entire technic of field work! Set several men to counting a pile of nails—some will count one at a time; some two at a time and some five at a time—but the complete tally will be the same no matter what system best fitted the idiosyncrasy of the counter! The best argument against meticulous precision or vast elaboration in the calculation is the admitted fact that the same Engineer cannot get the same results from the evaluation of the same scattered holes at two different times. But the percentage of disagreement will be well within safe limits. To be sure, there are occasional perplexing problems where the odd areas at the side are of unusual shape, or where it becomes necessary to separate the value of some particular tract or acreage from the calculation of the whole area. But these, too, yield to arithmetic and to the elementary principle of allowing each drill-hole to influence only that territory to which it is the nearest.

"HIGH HOLES"

Two or three of the holes are most likely to be very much higher than the others. Sometimes, they are so large in com-

parison with surrounding holes that there is a very natural suspicion of a grave error. There is no standard method of treating these "high holes" in the final calculation—different Engineers favor different methods. In fact, nowhere does the "personal" factor enter more strongly than in making the decision as to how such high holes shall be counted. Some hold it advisable to put down a check hole about six feet from the high hole. The check hole rarely shows values as great—sometimes it is surprisingly low. Usually, the average of the high hole and check hole is taken and figured into the final result as though there were but one hole at that point. It may be here noted that the gold content of more than one property has been padded by figuring in the check holes on an equal basis with the other drillings, the fact being forgotten or ignored that these holes were put down at favored locations and were not equally representative samplings. Usually, the Engineer is content to check one or two of the high holes and to regard the others in the light of what the first check holes reveal. If the property as a whole is of even enrichment, there is little likelihood that a second hole will confirm exceptionally high results of a first. On the other hand, if it is known to be "spotted", an occasional high hole can be accepted at full value without question. Indeed, the plotting of the several high holes will sometimes indicate a rich streak or channel. Extra holes are often put down to determine the course and extent of such an area of attractive concentration—not as check holes, but as supplementary drillings. Once that such an enrichment is established, there is of course no argument against accepting the higher values of the holes within such territory. If the high yield of a hole can be traced to a single large piece of gold, or a nugget, it is customary to leave out the weight of this persumably accidental bit in calculating the value. Some Engineers, in what is perhaps an over-anxiety for conservatism, do not include the high holes in their final calculations. Still others cancel them against the low holes—in the apparent conviction that the very high and the very low are due to errors which may be balanced. Another method is to take *half* the result of a high hole—which is more a concession to timidity than to common sense. Most dredge operators will report that a high hole is never proved up by the work of the dredge. Yet from actual results it seems to be an established and a safe and proper practice that the occasional high hole may be included in the calculations—if it was correctly drilled and if there is no mechanical error to be detected. For if the drill caught a stray bit of concentration behind a boulder or tapped a random streak of enrichment, there are quite certain to be many scattered areas of similar values that the dredge will recover. It seems fallacious and hypercautious to count out the one or two holes of most attractive result. But it must be remembered that only very rarely does such a high hole actually reveal a definite area of surrounding yardage of a corresponding richness—only adequate checking could prove that! But it does have a proper part in the summation of the whole!

CALCULATION OF STREAM CHANNEL VALUES

If the auriferous gravels follow a comparatively narrow and well-defined channel, it will have been prospected by lines of holes that cross it at intervals of a thousand feet, more or less, the holes spaced rather close together, depending on the width of the channel. The calculation of the values of the line may be made as is outlined and exemplified on Plates "A", "B," and "C." The principle is exactly the same as has been above outlined. Each drill-hole governs the length of "line" that is nearer to it than to any other drill hole. The figuring is facilitated by making a diagram of this channel cross-cut—being careful to lay out the dredgable limits at the side with due regard to the surface contour, bedrock elevation, barren gravels and practical digging depths. Ordinarily, the line will end at an actual drill-hole, but sometimes it will seem reasonable to extend it somewhat beyond a hole with acceptable values and depth part way to another hole that is either shallow or barren, or toward a further and approximated dredging limit. It is then necessary to credit to this "proportionate point" or "fictitious hole" a calculated value and depth—these assumptions to be made with due regard to the comparative distance of the two drill-holes that govern the estimate. This is shown on Plate "A"—there are such "proportionate points" at the right limit of Line 9 and at the left limit of Line 10. No matter if the holes be spaced irregularly, the area which each governs may be visualized and correctly calculated. To simplify calculations, the method exemplified in Plates "A" and "B" is often resorted to—the value of the area between two holes is estimated by multiplying the distance by the average depth and the value. A study of the Plates will reveal the whole process as reproduced from an actual report. To obtain the value of the "block", or the dredgable area between the two lines, the procedure as illustrated on Plate "D" may be followed. The calculation is sometimes called the determination of block values by the "Center of mass" formula. The lines are projected (by using the cosine of the angle which they make with an "altitude" line connecting their centers) and the average area of the projected bases, multiplied by this altitude, gives the volume of the "block." It should be noted that the lines affect the total value of the enclosed block not by their comparative *lengths*—but by the *area* of the channel that they cut. In the same way, the blocks may be added to find the total yardage and total gold content in a property—but the average value per cubic yard must be figured by allowing each block to affect the result in proportion to its yardage. Thus, the block yardages times the block value, added, divided by the sum of the yardages will give the value of the tract. It is often desirable to estimate as nearly as may be the value of a piece of land within certain property lines—often so bounded as to have only one or two holes actually within its confines. The calculation may be made by the use of "proportionate points" or fictitious holes—which are credited with a depth and a value that are derived from combining the value and depth of the two nearest holes in a proportion that is in inverse value to their distance away. Such estimates can be called little more than "intelligent guesses."

SUMMARY OF CALCULATION METHODS

In the discussion above, the plotting of the cross-cut of a channel has been pre-supposed. Referring to this drawing, the *Area* between two holes, which is ordinarily a trapezoid, is the average of their depth multiplied by the distance between them; the *Value* of this *Area* is the average value of the two holes multiplied by the *Area*; the *Value* of the *Line* is the sum of all these products divided by the *Area*;

the value of the *Block* is the sum of all these products in one *Line*, plus the sum of all these products in the second *Line*, divided by the added *Areas* of the two cross-sections, and multiplied by distance between the lines.

WHAT REPORT SHOULD COVER

The experienced Engineer will include in his final report of the property much beyond the mere calculation of yardage and values. For it is quite possible that that very report will be the dredge designer's sole source of accurate data regarding the physical conditions with which the dredge must cope. Modern firms prefer to design a dredge to fit the property. Even a used dredge, if torn down and transfered to a new property must be largely altered to meet the somewhat different conditions of soil and depth. Accordingly, there are here suggested the various bits of information upon which a thorough report will touch:

Character of bedrock.
Water-level with reference to surface and to bedrock.
Are boulders present? Their maximum size.
Character of gravel—loose, cemented, clay, etc.
Kind of gold—large or fine; flaky or rusty.
Are special gold-saving devices indicated—for nuggets or for flour or oxidized gold?
What are climatic conditions on property? Must dredge be specially equipped to lengthen its working season?
What are transportation conditions—with special reference to heavy dredge machinery?
What power is available for dredge construction and operation? Its probable cost?

The Panner at work

What is recommended size of dredge for efficient operation and for proper exploitation of available yardage?

What about water supply?

What is maximum grade of ground in which dredge will be called upon to work?

Where may dredge be best built with relation to greatest conveniences for construction and good values as well as in accord with the best economic plan for future operations?

The Mighty Dredge that follows in the path blazed by the KEYSTONE. This Dredge, on the Yuba River in California, is the largest in the world

CHAPTER IV
RELIABILITY OF KEYSTONE SAMPLINGS

Actual Instances of Comparison Between Drill Estimates and Dredging Returns

KEYSTONE ESTIMATES NOW CHECKED

If gold particles were uniformly distributed, uniform samplings would exactly determine the value of the whole. But the precious flakes are laid down according to no mathematical law; rather by the uncertainties of shifting currents, of spring freshets, of endless reconcentrations. Even a definite "pay-streak" begins unexpectedly and ends abruptly. Seldom will a dredge recover a comparative amount of gold in two successive days of operation. Yet the records of a generation of placer gravel exploration have proved that a reliable appraisal is quite possible with the Keystone Drill. Yet Engineers rarely expect a second drill-hole, sunk a few feet from a first, to yield values that are in close accord; nor are they surprised when a shaft, put down around a drill-hole, presents a considerable discrepancy in values. No one expects a flake of ore, chipped from one part of a vein, to assay the same as another flake from near-by. Yet the average of several such flakes will approximately reveal the proportion of mineral in the ore-body. The theory of *Keystoning* is the theory of averages—let the samples be fairly collected, correctly measured and properly assayed and they will collectively indicate the values of the whole. Once this was purely theory—today it is fact.

For in the last few years the results of vast dredging operations have been totaled and analyzed, and it is now possible to check them with the predictions of those who drilled the same areas in earlier years.

AGREEMENT ON OREGON PROPERTY

A placer deposit in Oregon which was the result of stream action was systematically drilled by lines of holes at right angles to the flow. The work was carefully done by competent Engineers and the calculations carried out by the methods illustrated in Plates "A" to "D." This property, consisting of 121 acres, has today been completely mined by a modern dredge. Its average depth was 18 feet and there was an overburden of almost barren tailings covering a "streaky" enrichment. The original estimates gave the average value per cubic yard as 16.8 cents and the value recovered by the dredge was 15.63 cents—or 93% of the estimate. The constant used in computing the drill results was .3333; had a constant of .3068 been used, the estimated and actual values would have agreed quite closely. There were here put down one hole to every 2.4 acres.

Yet hardly any two of the eight "blocks", into which the area was divided by the drill lines yielded the identical values which the drill forecast—in one of them the recovered value per cubic yard exceeded the estimate by 49.4% and in another it was less by 68.2%.

CHECK ON A CALIFORNIA PROPERTY

We reprint from the Engineering and Mining Journal an account of the balancing of prospecting estimates and dredging results on a California property section of 118.5 acres where the results of 38 drill-holes were used in making the forecast—14 of the holes being outside but adjacent to the area dredged.

"The estimated value per cubic yard in this tract was 29.88 cents and the dredge returns were 31.55 cents, a gain of 5.6%. We believe the constant used in figuring the drill returns on this property to have been .3068. Some years ago on another and undredged portion of this property consisting of 493 acres an estimate was made to determine the gold content. The results from 53 drill holes, which were within the limits of the tract, were used by taking 70 per cent of their recorded value. Consideration was then given and use made of the values obtained from dredging operations which had been conducted adjacent to and around about two thirds of the tract. Since then 402 acres of this tract have been dredged and the returns per cubic yard averaged 12.73 cents. The data used in making the original estimate were applied to the portion now dredged and the proportionate values found to be 12.70 cents per cubic yard."

ANOTHER CHECK FROM LARGE ACREAGE

From the same article we quote another example that shows a recovery closely corresponding to the estimate, giving the results produced by operations on three separate tracts in a large California property, the third tract having been mined by three dredges:

Tract	Average Depth	Acres	No. of Holes	Acreage to Each Hole	*Value per Cubic Yard* Drill	Dredge Recovery	Per-centage
A	22.5′	173.5	x 57	3.2	6.8	7.82	115%
B	44.5′	84.0	20	4.2	5.9	6.7	113%
C Dr. #1	51.8′	183.0	120	1.5	11.1	9.64	87%
C Dr. #2	60.6′	106.0	41	2.6	11.2	9.44	84%
C Dr. #3	56.4′	135.0	58	2.3	11.6	11.30	97%

Moving under difficulties with the Non-Traction

x 37 shafts and 20 Drill-holes
On Tracts A and B Constant Used in Calculating Drill Results .27
On Tract C Constant Used in Calculating Drill Results .30

CHECK FROM SMALL DREDGE OPERATIONS

"The average results of all the above, proportioned to the acreage, is a drill value of 9.48 cents per yard, and the dredge recovery 9.12 cents, or 96.2%".

And we quote yet another instance:

"Dredging operations on one property in California have produced returns which, taken as a whole, correspond quite closely with the original estimates, and the following details have been presented covering the work done in the past three years:

Year	Average Depth	Acres	No. of Holes	Acreage to Each Hole	*Value per Cubic Yard* Drill	Dredge Recovery	Per-centage
1918	32.1′	19.94	11	1.8	10.39	10.64	102.4%
1919	34.4′	20.90	10	2.1	9.69	9.22	95.2%
1920	29.8′	20.43	7	2.9	10.69	14.34	134.0%

EXAMPLE FROM OPERATIONS OF NATOMAS CONS. OF CALIFORNIA

"The average results proportioned to the acreage show an estimated value of 10.25 cents and a dredge recovery of 11.39 cents, or an increase of 11.1%".

Perhaps the dredging problems of the Natomas Company of California have been the most difficult encountered on large scale operations. Here, on the American River, are high benches of gravel that is so compact and cemented as almost to defy the dredge buckets and, down in the bottom lands, wide areas of softer and shallower ground. A check, made five years ago on the operation of three of their many dredges gives a reasonably close agreement.

Dredge	Area in Acres	Years of Operation	Percentage of Estimate Recovered
No. 5	180	10	86
No. 8	106	4	87
No. 9	135	6	101

These results are based on an original calculation using the constant .30. During the last five years these dredges, as a whole, have produced about 101 per cent of the calculated recovery!

MONTANA PROPERTY

A portion of a Montana property consisting of 300 acres was prospected with 77 drill holes spaced at irregular distances. The average value per cubic yard indicated by the drilling was 15.83 cents, and the dredge recovery 13.55 cents or 85.6%. The average depth of the ground was 40 feet and the larger portion of the values was contained in the three feet of gravel next to the bedrock. Such a deposit offers obvious difficulties to accurate calibration.

TABULATION OF AVAILABLE COMPARISONS

The article from the Engineering and Mining Journal partly reproduced above, written by Mr. Charles W. Gardner, Manager of the Mines Operating Department of the Hammon Engineering Company, probably the first Engineer to use a Keystone Drill for placer prospecting, concludes with a summary that is of striking interest:

"From all of the properties above mentioned we are able to segregate 3,743 acres to which we can apply data given in fairly accurate reports. This combined area was prospected by means of 1,749 drill-holes, or one to every 2.1 acres. The average value per cubic yard obtained by drilling was 15.4 cents and the average dredge recovery 13.55 cents, or 88%. A comparison of returns, segregated as to values, is as follows:

Drill Values per Cubic Yard	*Average Value per Cubic Yard*				*Percentage*	
	Acres	Per Centage of Whole	Drill Value	Dredge Recovery	Gain	Loss
Under 11¢	27.9	0.7%	9.21¢	8.60¢		6.6%
Under 11¢	1638.1	43.8	7.68	9.34	21.6%	
Between 11 and 12¢	424.0	11.3	11.28	10.12		10.3
Between 11 and 12¢	480.0	12.8	11.61	16.44	41.6	
Between 12 and 20¢	582.5	15.6	16.92	13.10		22.6
Over 20¢	392.0	10.5	45.86	20.93		54.4
Over 20¢	198.5	5.3	33.15	35.97	8.5	
Totals	3743.0	100.0%				
AVERAGES			15.40	13.55		12.0

Here is spread before us the work of different Engineers at different times in many fields. There has been no attempt to trace out the avoidable error in either prospecting or mining. Suffice it to say that there is included in this tabulation the exploration of at least two properties where the excavated material was a clean washed sand and gravel with little clay and many boulders; making the forecasting of the values exceedingly difficult by any method—and the recovery of certain dredges that fall short of the highest standard of modern efficiency.

ACCURATE PROSPECTING POSSIBLE

Mr. Gardner closes his article with; "We feel that a safe conclusion to be drawn from all of the above is that when a property has been sufficiently prospected by an experienced Engineer, the results intelligently interpreted and the calculations accurately made, the result obtained will indicate within reasonable limits the gold content.

Belmonte, Portugal, First KEYSTONE in Portugal, 1912

"Full consideration should be given all these points and then if the estimated gross value safely and sufficiently exceeds the cost of acquisition, equipment and operation, there should be nothing to deter or discourage the investment of capital in such an enterprise."

HOW DIFFERENT CONDITIONS AFFECT ACCURACY

Mr. James W. Neill, in an article in the Mining and Scientific Press, says, "If the driller keeps his bit behind his shoe, and sees that he gets a correct amount of core, there should be little question of the correctness of a *large general average,* and the Engineer can use such factors of safety as his experience and the character of the ground indicate. In very loose ground, I personally look for a recovery of full drill values where the gold is coarse and is entirely contained in the foot or two above bedrock. I think one will also usually overrun the drill, provided the bedrock can be dug. That is about the sum of our experience at Snelling."

We quote an extract from an article by the compiler of this booklet that recently appeared in the Engineering and Mining Journal; "It may be observed that the accuracy of results gained by drilling is surprising, considering the comparatively small size of the sample. For example, one hole to two acres in 50-ft. ground would mean that 1/328,000 part of the gravel was examined. Of course one hole by itself means nothing—it is only a carefully charted series that reliably represents actual conditions.

"All available data seems to point to the following facts regarding the accuracy of drilling:

1. Drill results give high assays where the gold is fine.

2. Estimates will tally with returns where gold is reasonably heavy; ground is compact with a little clay; and prospecting and dredging are intelligently done.
3. The drill will exaggerate the value of *very* loose ground where sand and water pressure crowd material into the bottom of the pipe.
4. The drill will signally fail to show recoverable gold in areas of moderately loose ground where gold is distributed and in comparatively small amount."

SUMMARY

The whole problem is squarely up to the discretion of the Engineer. He may compensate for the co-efficient of the personal accuracy of his crew; he may use the proper constant in his calculations; he may use his own experience and the now recorded findings of others—and translate the results of his Keystone Drill work to a forecast that bears the stamp of conviction and accuracy!

CONCLUSION

The Keystone Driller Company sells more than a mere machine—it sells service! Orders are carefully filled with selected materials and shipments forwarded with all possible expedition for we are keenly alive to the difficulties and hazards of field work. During the score of years in which Keystone Drills have been the reliance of prospectors the world over, we have collected a considerable knowledge of the technic of placer testing and of the proper equipment therefor. That data we shall gladly share; and we offer our experience to any Engineer, be he outfitting for a proposed campaign of prospecting or meeting a knotty problem in the field!

PLATE "A"

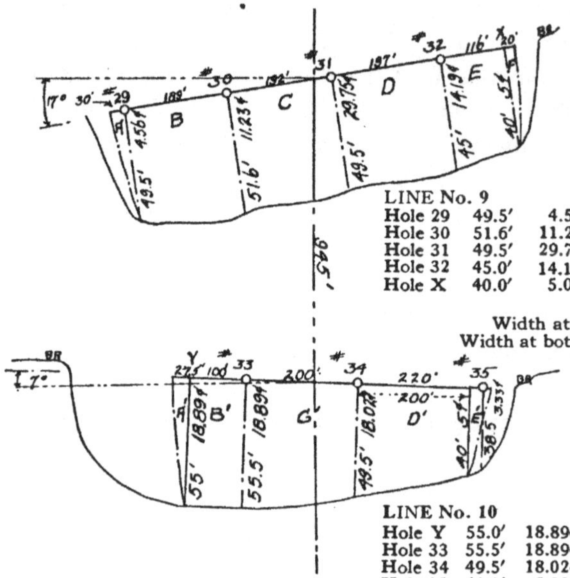

LINE No. 9

Hole 29	49.5′	4.56c	30′
Hole 30	51.6′	11.23c	189′
Hole 31	49.5′	29.75c	192′
Hole 32	45.0′	14.19c	197′
Hole X	40.0′	5.00c	116′
			20′
		Width at top	744′
		Width at bottom	694′

LINE No. 10

Hole Y	55.0′	18.89c	27.5′
Hole 33	55.5′	18.89c	100′
Hole 34	49.5′	18.02c	200′
Hole 35	40.0′	5.00c	200′
			20′
		Width at top	547.5′
		Width at bottom	500′

KEYSTONE CREEK PLACERS

(From actual drill records)

Key Chart to Estimate of Line 9 on Plate "B"; of Estimate of Line 10 on Plate "C"; of Block Estimate on Plate "D."

NOTE

Left limit of Line 9 determined by low value actual drill-hole. Right limit of Line 9 determined by approximation of dredging limit—and marked by "fictitious" drill-hole "X."

Left limit Line 10 determined by some physical factor—probably an irregular rim. "Fictitious" Hole "Y." Right limit Line 10 determined by too low value of Hole #35. Engineer has approximated limit of values 20 feet inside of this Hole—and marks "fictitious" hole #35′ giving it a limiting value of 5¢.

PLATE "B"

KEYSTONE CREEK PLACERS—Valuation of Line 9

Area of Block		Square Feet		Cubic Yards		Cents		
A	equals $\frac{49.5 \times 30}{2}$	742.5	or	27.5	x	3.04	=	83.6
B	equals $\frac{49.5 \,\&\, 51.6}{2}$ x 189	9553.9	or	353.8	x	7.89	=	2791.4
C	equals $\frac{51.6 \,\&\, 49.5}{2}$ x 192	9705.6	or	359.5	x	20 49	=	7366.1
D	equals $\frac{49.5 \,\&\, 45}{2}$ x 197	9308.2	or	344.8	x	21.97	=	7575.2
E	equals $\frac{45 \,\&\, 40}{2}$ x 116	4930.0	or	182.6	x	9.59	=	1751.1
F	equals $\frac{40 \times 20}{2}$	400.0	or	14.8	x	3.33	=	49.3
		34640.2		1283.0		15.25		19616.7
	One foot into bedrock	694.0						
		35334.2		1308.7		14.99		19616.7

NOTES

There is given to Areas "A" and "F" a value equal to only two thirds of their governing drill-hole. These small areas are the approximation of the departure from the vertical of the dredge bank.

Under the "Cents" column is the average value of the two bounding drill holes.

To the yardage total has been added the barren material resulting from dredging one foot into bedrock.

The figure 14.99 represents the value in cents per cu. yd. of the line and is gained by dividing 19616.7 by 1308.7.

PLATE "C"

KEYSTONE CREEK PLACERS—Valuation of Line 10

Area of Block		Square Feet		Cubic Yards		Cents		
A	equals $\frac{55 \times 27.5}{2}$	756	or	28.0	x	12.59	=	352.5
B	equals $\frac{55 \& 55.5}{2} \times 100$	5525	or	204.6	x	18.89	=	3864.9
C	equals $\frac{55.5 \& 49.5}{2} \times 200$	10500	or	389.0	x	18.46	=	7180.9
D	equals $\frac{49.5 \& 40}{2} \times 200$	8950	or	331.5	x	11.51	=	3815.5
E	equals $\frac{40 \times 20}{2}$	400	or	14.9	x	3.33	=	49.7
		26131		968.0		15.72		15263.5
	One foot into bedrock	500		18.4				
		26631		986.4		15.48		15263.5

NOTES

See Notes on Plate "B."

The figure 15.48 represents the value in cents per cu. yd. of the line and is gained by dividing 15263.5 by 986.4.

PLATE "D"

KEYSTONE CREEK PLACERS

Estimate Gross Value of Block Between Lines 9 and 10

Area of Section	Projected Area	Gross Value	Cubic Yards
Line 10 = 26131 x .9925 (Cos. 7°) =	25935	15263.5	968
Line 9 = 34640.2 x .9563 (Cos. 17°) =	33126	19616.7	1283
	2)59061	34880.2	2251
	29530.5		

29530.5 (Average area projected base) multiplied by 945 (Surveyed or plotted distance between middle of both lines which is altitude of "block") gives a total cubic content of 27,906,522 Cubic Feet.

27,906,522 Cu. Ft. equals 1,033,367 Cu. Yds.

The Gross *Value* Factor, divided by *Cu. Yds.* Factor equals:

2251)34880.2 (15.49 cents — Value of block

Total Value of Block equals 1,033,367 times 15.49 or $160,099.70.

As a check on Calculations we may figure the same Total by using the products of the Line Calculations after adding "One Foot Into Bedrock."

Area of Section	
Line 10 = 986.4 Cu. Yds. x .9925 (Cos. 7°) =	979.0
Line 9 = 1308.7 Cu. Yds. x .9563 (Cos. 17°) =	1251.5
	2)2230.5
	1115.25

1115.25 x 945 = 1,053,911 Cu. Yds.

Total Value of Block equals 1,053,911 x 15.19c = $160,089.08.

NOTE

A line is drawn connecting the centre of the Lines 9 and 10. These lines are then projected so as to form a 90° angle with this "altitude" line. The projected base areas are added, averaged, and multiplied by the altitude to find content of block in Cu. Yds. Additions and subtractions might have been made for irregular rim.

PLATE "E"

CORE MEASUREMENTS—KEYSTONE PLACERS

			Core Measured in Pipe			Core Measured on Surface Without Slimes		With Slimes		Total Cubic Feet as per
Hole No.	Depth Ft.	Drive Ft.	Total Ins.	Per Ft. Ins.	Total Cu. Ft.	Total Cu. Ft.	Average Cu. Ft.	Total Cu. Ft.	Average Cu. Ft.	Const. .27
1	110	110	1446	14.32	21.69	30.27	.2752	(Not Measured)		29.70
2	105	98.5	1539	15.83	23.08	25.18	.2556	39.53	.401	26.60
3	105	98.5	1521	15.44	22.82	26.30	.2669	37.80	.384	26.60
4	105	97.6	1477	15.14	22.16	24.87	.2547	38.82	.398	26.35
5	105	96.1	1600	16.65	24.00	26.33	.2740	38.02	.396	25.95
6	105	97.4	1493	15.33	22.40	24.35	.2499	36.00	.369	26.30
7	107	99.7	1529	15.34	22.94	22.36	.2243	28.75	.288	26.91
8	98.5	83.0	1402	16.90	21.03	25.24	.3041	32.30	.389	22.41
	840.5		12007	15.34	180.12	204.90	.2621	251.22	.3746	210.81

NOTES

Figures under "Depth" represent total from surface—under "Drive" they represent actual drilling. (Holes were started in a pit.)

This Chart, taken from the Report of an Engineer, was prepared by him that he might compare the core records of various similar holes. It shows work that is reasonably uniform, with no outstanding discrepancies, and proves that the material actually recovered, if discounted for the "swelling" of the gravels, checks very closely with the theoretical displacement of the drive-shoe.

PLATE "F"
KEYSTONE DRILLER COMPANY
TABLE FOR FINDING VALUE PER CUBIC YARD

(When depth of drill-hole and weight of gold in milligrams are known)

CONSTANT .27

Depth to Bedrock in Feet	Factor	Depth to Bedrock in Feet	Factor	Depth to Bedrock in Feet	Factor
10	.6000	26.5	.2264	43	.1395
10.5	.5714	27	.2222	43.5	.1379
11	.5455	27.5	.2182	44	.1364
11.5	.5217	28	.2143	44.5	.1348
12	.5000	28.5	.2105	45	.1333
12.5	.4800	29	.2069	45.5	.1319
13	.4616	29.5	.2034	46	.1305
13.5	.4444	30	.2000	46.5	.1290
14	.4286	30.5	.1967	47	.1277
14.5	.4138	31	.1935	47.5	.1263
15	.4000	31.5	.1905	48	.1250
15.5	.3871	32	.1875	48.5	.1237
16	.3750	32.5	.1846	49	.1224
16.5	.3636	33	.1818	49.5	.1212
17	.3529	33.5	.1791	50	.1200
17.5	.3429	34	.1765	50.5	.1188
18	.3333	34.5	.1739	51	.1177
18.5	.3243	35	.1714	51.5	.1165
19	.3158	35.5	.1690	52	.1154
19.5	.3077	36	.1667	52.5	.1143
20	.3000	36.5	.1644	53	.1132
20.5	.2927	37	.1622	53.5	.1121
21	.2857	37.5	.1600	54	.1111
21.5	.2791	38	.1579	54.5	.1101
22	.2727	38.5	.1559	55	.1091
22.5	.2667	39	.1539	55.5	.1081
23	.2609	39.5	.1519	56	.1071
23.5	.2553	40	.1500	56.5	.1062
24	.2500	40 5	.1481	57	.1053
24.5	.2449	41	.1463	57.5	.1044
25	.2400	41.5	.1446	58	.1035
25.5	.2353	42	.1429	58.5	.1026
26	.2308	42.5	.1412	59	.1017

Diameter of Cutting-Shoe 7½ inches. Value of gold 0.06 cents per milligram—$18.66 per ounce. To find value in cents per cubic yard, multiply number of milligrams of gold recovered from drill-hole by the factor in the table opposite the depth to bedrock

PLATE "G"
KEYSTONE DRILLER COMPANY
TABLE FOR FINDING VALUE PER CUBIC YARD

(When depth of drill-hole and weight of gold in milligrams are known)

CONSTANT .3068

Depth to Bedrock in Feet	Factor	Depth to Bedrock in Feet	Factor	Depth to Bedrock in Feet	Factor
10	.5280	26.5	.1992	43	.1228
10.5	.5029	27	.1955	43.5	.1214
11	.4800	27.5	.1920	44	.1200
11.5	.4591	28	.1886	44.5	.1187
12	.4400	28.5	.1853	45	.1173
12.5	.4224	29	.1821	45.5	.1160
13	.4061	29.5	.1789	46	.1148
13.5	.3911	30	.1760	46.5	.1136
14	.3771	30.5	.1731	47	.1123
14.5	.3641	31	.1703	47.5	.1112
15	.3520	21.5	.1676	48	.1100
15.5	.3407	32	.1650	48.5	.1089
16	.3300	32.5	.1625	49	.1078
16.5	.3200	33	.1600	49.5	.1067
17	.3106	33.5	.1576	50	.1056
17.5	.3017	34	.1553	50.5	.1046
18	.2933	34.5	.1530	51	.1035
18.5	.2854	35	.1509	51.5	.1025
19	.2779	35.5	.1487	52	.1015
19.5	.2707	36	.1467	52.5	.1006
20	.2640	36.5	.1447	53	.0996
20.5	.2575	37	.1427	53.5	.0987
21	.2514	37.5	.1408	54	.0978
21.5	.2456	38	.1389	54.5	.0969
22	.2400	38.5	.1371	55	.0960
22.5	.2347	39	.1354	55.5	.0951
23	.2296	39.5	.1337	56	.0943
23.5	.2247	40	.1320	56.5	.0935
24	.2200	40.5	.1304	57	.0926
24.5	.2155	41	.1288	57.5	.0918
25	.2112	41.5	.1272	58	.0910
25.5	.2071	42	.1257	58.5	.0903
26	.2031	42.5	.1242	59	.0895

Diameter of Cutting-Shoe 7½ inches. Value of gold 0.06 cents per milligram—$18.66 per ounce. To find value in cents per cubic yard, multiply number of milligrams of gold recovered from drill-hole by the factor in the table opposite the depth to bedrock.

PLATE "H"
KEYSTONE DRILLER COMPANY
TABLE FOR FINDING VALUE PER CUBIC YARD
(When depth of drill-hole and weight of gold in milligrams are known)
CONSTANT .33⅓

Depth to Bedrock in Feet	Factor	Depth to Bedrock in Feet	Factor	Depth to Bedrock in Feet	Factor
10	.4860	26.5	.1834	43	.1130
10.5	.4629	27	.1800	43.5	.1117
11	.4418	27.5	.1767	44	.1105
11.5	.4226	28	.1736	44.5	.1092
12	.4050	28.5	.1705	45	.1080
12.5	.3888	29	.1676	45.5	.1069
13	.3738	29.5	.1647	46	.1057
13.5	.3600	30	.1620	46.5	.1045
14	.3471	30.5	.1593	47	.1034
14.5	.3351	31	.1568	47.5	.1023
15	.3240	31.5	.1543	48	.1013
15.5	.3135	32	.1518	48.5	.1002
16	.3037	32.5	.1495	49	.0992
16.5	.2945	33	.1472	49.5	.0982
17	.2895	33.5	.1450	50	.0972
17.5	.2777	34	.1429	50.5	.0963
18	.2700	34.5	.1408	51	.0953
18.5	.2627	35	.1388	51.5	.0944
19	.2558	35.5	.1369	52	.0935
19.5	.2492	36	.1350	52.5	.0925
20	.2430	36.5	.1331	53	.0917
20.5	.2371	37	.1313	53.5	.0909
21	.2314	37.5	.1296	54	.0900
21.5	.2260	38	.1279	54.5	.0892
22	.2209	38.5	.1262	55	.0884
22.5	.2160	39	.1246	55.5	.0876
23	.2113	39.5	.1230	56	.0868
23.5	.2068	40	.1215	56.5	.0860
24	.2025	40.5	.1200	57	.0853
24.5	.1984	41	.1185	57.5	.0845
25	.1944	41.5	.1171	58	.0838
25.5	.1906	42	.1157	58.5	.0831
26	.1869	42.5	.1143	59	.0824

Diameter of Cutting-Shoe 7½ inches. Value of gold 0.06 cents per milligram—$18.66 per ounce. To find value in cents per cubic yard, multiply number of milligrams of gold recovered from drill-hole by the factor in the table opposite the depth to bedrock.

PLATE "I"

KEYSTONE DRILLER COMPANY
GOLD DREDGE CAPACITY CHART

Cubic Yards Elevated per Month of 30 Days at
Various Efficiencies for Different Sizes
(Bucket Speed at 50 Feet per Minute)

Figures are based on an actual operating time of 20 hours out of each 24.

Size of Dredge	100% Efficiency	80% Efficiency	60% Efficiency
2½ Cu. Ft.	89,000	71,000	53,000
3½ Cu. Ft.	108,000	86,000	65,000
5 Cu. Ft.	140,000	112,000	82,000
7½ Cu. Ft.	182,000	146,000	110,000
9 Cu. Ft.	213,000	171,000	129,000
15 Cu. Ft.	304,000	244,000	184,000

From this table the Engineer may roughly approximate the yardage of a proposed dredge—once he knows the physical and climatic conditions to be encountered and general working conditions.

PLATE "J"

YARDAGE CALCULATION TABLE

DREDGE YARDAGE TABLE

Acres

Depth	10	20	30	40	50	60	70	80	90	100
10	161,333	322,667	484,000	645,333	806,667	965,000	1,129,333	1,290,667	1,452,000	1,613,333
20	322,667	645,333	968,000	1,290,667	1,613,333	1,936,000	2,258,667	2,581,333	2,904,000	3,226,667
30	484,000	968,000	1,452,000	1,936,000	2,420,000	2,904,000	3,388,000	3,872,000	4,356,000	4,840,000
40	645,333	1,290,667	1,936,000	2,581,333	3,226,667	3,872,000	4,517,333	5,167,667	5,808,000	6,453,333
50	806,667	1,613,333	2,420,000	3,226,667	4,033,333	4,840,000	5,646,667	6,453,333	7,260,000	8,066,667
60	968,000	1,936,000	2,904,000	3,872,000	4,840,000	5,808,000	6,776,000	7,744,000	8,712,000	9,680,000
70	1,129,333	2,258,667	3,388,000	4,517,333	5,646,667	6,776,000	7,905,333	9,034,667	10,164,000	11,293,333
80	1,290,667	2,581,333	3,862,000	5,162,667	6,453,333	7,744,000	9,034,667	10,325,333	11,616,000	12,906,667
90	1,452,000	2,904,000	4,356,000	5,808,000	7,260,000	8,712,000	10,164,000	11,616,000	13,068,000	14,520,000
100	1,613,333	3,226,667	4,840,000	6,453,333	8,066,667	9,680,000	11,293,333	12,906,667	14,520,000	16,133,333

To find acreage in any given Track:
For Example—120 Acres at 35′-0″ Depth.

100 Acres at 30′-0″ Depth =	4,840,000
100 Acres at 5′-0″ Depth—Using 1/10 of 50′-0″ Depth =	806,667
20 Acres at 30′-0″ Depth =	968,000
20 Acres at 5′-0″ Depth—Using 1/10 of 50′-0″ Depth =	161,333
27	
Total	6,776,000

Proof: $\frac{120 \times 35 \times 43560}{27} = 6,776,000$

FIELD LOG

No. of Hole..............

Date Commenced.......................... Date Finished..........................

DEPTH FEET	No. of COLORS Size 1	Size 2	Size 3	FORMATION	CORE	REMARKS

Total Depth........feet. Bedrock........feet. Water level........feet.

ABBREVIATIONS

T.	Tailings	S.	Sand	St.	Sticky
F.	Fine	G.	Gravel	Md.	Medium
C.	Coarse	Cl.	Clay	M.	Much
V.	Very	Sm.	Some	L.	Loose

PART II

AUTHORITATIVE ARTICLES ON MINERAL PROSPECTING

Reprinted from the Keystone Driller Company's Catalog No. 2, Edition 1907, and from various technical Journals.

THE PROSPECTING AND VALUING OF DREDGING GROUND

Written for the Mining and Scientific Press
By Norman C. Stines

[Published in the Issues of Feb. 3 and Feb. 10, 1906.]

The prospecting of gravel deposits to test their fitness for dredging purposes is done by boring holes with a drilling machine, by sinking shafts, or by making a trial run with what is known as a prospecting dredge. It is such work by means of the Keystone drill that this article attempts to describe.

In prospecting a piece of ground, the following data must be ascertained:

1. The average value of the ground per cubic yard and the distribution of the gold.

2. The character of the gravel.

3. The character of the bedrock and its approximate contour.

4. The position of the water level as referred to the surface.

5. The amount of water obtainable for the pond.

6. The nature and cost of the power obtainable.

7. The length of the working season.

8. The cost of the land.

The first four points are determined directly by the use of the drill, and the last four by a careful study of the prevailing conditions. As this paper is to describe the method of using the drill and the interpretation of the results thus obtained, it will treat of the first four points only.

The machine used in the greater part of this work in California is the Keystone Driller No. 3, Traction. It is made by the Keystone Driller Company, Beaver Falls, Pa., and costs (at present, 1923) complete f. o. b. at the factory, about $2700.00. The driller consists of a walking-beam arrangement operated by a steam engine of 11 h.p. This produces the required motion for raising and dropping the drill, and for transporting it from place to place. The engine, with the drilling cable reel, the sand-reel and the socket, drill-stem and bit, is suspended by a 1¾″ cable which passes over the rear sheave and over the front one, continuing over the sheave at the top of the derrick and then down. As the walking beams are put in motion, the drill is alternately raised and dropped. The machine makes about 52 strokes per minute in drilling and about 54 in driving.

LAYING OFF THE GROUND

For the preliminary work, to determine if the gold is scattered over a large area or if it is confined to a narrow winding channel, a few holes are sunk from 500 to 700 ft. apart. If the gold is found to be pretty evenly distributed over the whole tract, it is divided into five or ten-acre squares, according to the amount of drilling to be done. A flag is placed in the center of each of these squares, and this marks the site of the hole.

If the gold is found to be confined to a narrow winding channel, the ground is crossed, at right angles to the channel, by series of holes. These series are from 400 to 800 ft. apart. At every 100 ft. in each series a flag is placed and this, as

Changing Tools

before, marks the place to be drilled. On drilling, the first hole in any series is placed as near the center of the channel as possible, and then drilling to one side or the other is only carried as far as the results warrant. Every fourth or fifth series is drilled across the tract to be sure that there are no splits in the channel. This method insures the least amount of work spent in valueless ground. The series are designated by letters and the holes by numbers. In this way they are recorded in the transit notes. When a hole is drilled, the number in the log-book is also placed in the transit book and the number from the transit book is placed in the log-book.

OPERATION OF DRILLING THE HOLE

The machine is moved to one of the flags and set up there. A hole is then dug to a couple of feet and the shoe-joint is dropped, then plumbed, and finally the dirt filled around it. The next operation varies according to the nature of the ground. Where the soil is deep, the driving blocks are put on and the casing is driven as far as possible, or to the gravel. Where the gravel commences at the surface, the drill is first lowered and the ground immediately beneath the shoe is drilled. The casing generally settles as the drilling proceeds; but if it does not, when a sufficient depth has been drilled, the blocks are put on and the casing driven to a point about three inches below the depth drilled. The lower end of the shoe-joint is protected from injury by a steel shoe, 7½″ in diameter, tempered at the cutting edge and slightly beveled on the inside. It is the area of this shoe which represents the area excavated.

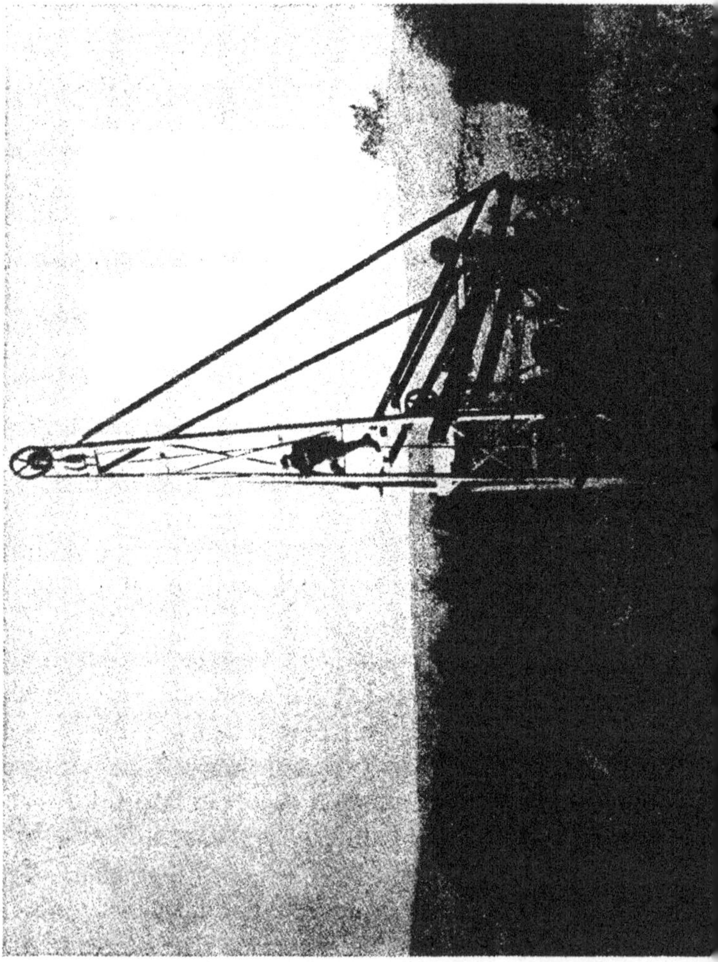

The casing is nearly always driven to gravel. If the first joint does not reach, a second, and sometimes a third, is put on. As soon as the first joint has been driven to head, the driving cap is removed, the threads brushed off and greased (being careful to use only graphite and linseed oil for the purpose so as not to allow any oil to get into the bore hole, K. D. Co.). The second joint is then put on, cinched tight, the driving cap placed on that and the driving continued. The depth driven at first usually varies from three to twelve feet.

In driving, the stem on which the driving blocks have been placed acts as the weight in a pile-driver. The weight of the stem is about 800 lbs. and the drop ranges from 28 to 36 inches.

After the casing has been driven as far as desired, the driving blocks are removed, the stem is lowered into the casing, the core is measured, water is poured in and the walking beams are set in motion. When the machine man measures and reports the length of core to the panner, he also reports the depth of the casing in the ground. The core is measured in the following manner: The length of connected casing is measured on the drill-stem and rope, commencing at the bit. This point is marked in chalk. The stem is then lowered into the hole, and the difference between where the top of the casing is and where the mark on the stem is, is the core. The core then is the material in the casing. This is important and will be fully explained later.

While there is yet some core in the casing, the drill-stem is removed and the sand-pump is placed in the hole. This is a piece of 4-in. pipe about 8 ft. long, in which a rod or plunger fits the pipe closely. At its lower end there is a valve. On raising the plunger suddenly, a vacuum is formed and the material in the hole is forced in through the valve at the bottom. This is repeated until all the material is thought to be in the pump. The pump is then reeled out of the hole, and its contents is caught by the panner. The stem is then lowered into the casing, the core is measured, the stem is hoisted, the driving blocks are put on and the casing is driven again.

This is repeated for each foot of the hole until the desired depth has been reached. The desired depth is generally bedrock or the lower limit of a known pay-streak. The depth to which the hole is sunk in bedrock depends upon the occurrence of gold. Some bedrock carries gold for a considerable depth. The hole is generally put down until no colors are found in the pan. It may happen that the casing has come on bedrock in such a way as to cross a seam. This seam may be particularly rich and gold might come into a number of pans after bedrock has been reached. When this occurs, it is well to be wary of the results of that hole if it gives a yield above the general average. The gold is plainly coming from a seam and so we have no measure of the volume of dirt from which it is coming. There is one way to guard against this danger. When approaching bedrock, use two or more small pans for receiving the concentrate from each foot panned. The concentrate for, say, the first 27 ft. is placed in a pan. On driving the 28th foot and panning, there is found to be more gold than usually comes with bedrock. This is then put in a separate small pan, as is each succeeding pan of concentrate; if there seems to be more gold than usual, it is saved separately and the results of the ground above bedrock and that below bedrock are then known. This is not often necessary, but it will save distrust of the high returns, which sometimes appear in a hole among a lot of other lower results, which are fairly close together.

The hole being drilled as far as desired, the drill is placed on the ground, the rope-socket taken from it and placed on the pipe-jars. These pipe or pulling-jars consist of an iron ram or boss on the end of a stem about four feet long. On this stem is a threaded knocking-head with a square opening through which the stem passes, up and down, in striking the blow. The threaded knocking-head is screwed to the casing and the ram is drawn up so that it will strike against the knocking-head when put in motion by the walking-beams. The walking-beams are then set in motion and the jar of the ram against the pipe causes it to be loosened and to be drawn up. As the pipe comes up, the slack is taken by the runner. When the top joint is removed from the ground, the machine is stopped, the threaded knocking-head is removed and the pipe-jars are pulled from the casing. The top joint is then removed and the operation is repeated for each succeeding joint.

When all of the joints are out of the ground, the drill-stem is again put on the rope-socket, loaded into the bed of the machine, the jacks removed from under the machine, the engine thrown into gear for propelling mechanism and the whole is moved to the next flag.

TREATMENT OF MATERIAL FROM THE HOLE

The material brought up by the sand-pump may be treated in different ways. It may be caught in a pan held over the sluice-box, the slime going to the sluice-box; it is then washed in the panning tub. It may be caught in the pan, the slime going to the sluice-box as before, then washed in a rocker, the material caught on the apron being panned separately; or it may be dumped into the sluice-box and all of it swept into the rocker, the concentrate being panned.

The second way seems to answer the purpose best. It gives quickest results and is the easiest for the panner. It uses a minimum of water, and at all times the water in the tub is comparatively clear, thereby preventing the loss of gold unavoidable when panning in thick water.

The apparatus required is a sluice-box about eight feet long and 12 in. square in cross-section, a small rocker, pans and panning tub.

Each foot is pumped and panned; the gold is estimated carefully. The amount is then inserted on the log opposite the foot from which it came. This is more fully explained when describing the log-book. The gold is classed in three sizes, 1, 2 and 3. Number 3 is the finest and consists of all pieces which weigh less than one milligram. Number 2 gold would be any piece weighing one milligram or over, up to four milligrams, and Number 1 is such that any piece weighs over four milligrams.

As mentioned above, the concentrate from each washing is put in a small pan, and, when the hole is completed, the gold is amalgamated and put by itself. In special cases, we would have more than one "pill", as explained already. The gold is separated from the mercury by nitric acid and thoroughly washed, dried, annealed and weighed. If the drying is done in the annealing cup, there is no danger of loss due to sputtering, but the addition of a few drops of alcohol to the last wash water, when using the porcelain cups, prevents sputtering. From the weight of gold obtained, and the volume of dirt from which it came is figured the value of the ground immediately adjacent to the hole.

Not only the gold obtained from the pannings is recorded, but the character of the gravels. This would include a note on the size of particles and whether there was sand, clay or cementing material present. This is ascertained by a close examination of the screenings, as left in the hopper of the rocker. Then, too, the panner carefully notices his concentrate and records anything of interest, such as the amount of black sand, any gem stones, and the appearance of amalgam. These are all recorded in the log-books, opposite the foot in which they were found.

THE LOG BOOK

It is in the records that the engineer finds his data for valuing ground. The log-book as kept by different men varies, but for practical utility I have found the accompanying to be best; Table I is a page from a log-book as kept by a panner. All linear dimensions are in feet and tenths. At the head of the page is placed the number of the hole as drilled and the number as recorded in the transit-book. Following this is the name of the tract on which the hole was drilled. In column "A", we have recorded the depth to which the casing has been driven. Everything on the same horizontal line as that refers to that foot in the section of the ground. This is measured by the machineman and recorded by the panner. In column "B", we have the depth of core after driving. This is a measure of the amount of material which has been forced into the pipe by the last drive *plus* the core left in the pipe before driving. It gives the panner an idea of the kind of ground to expect and occasionally it indicates the character of bedrock. In drill-ground in which the bedrock is volcanic ash (such as that at Oroville), a large core after a hard drive is a very good sign of bedrock. Its importance, however, lies in the fact that it acts as a check on the amount of material coming into the pipe. Column "C" is also important. In it we can see if the bit has been below the casing and at what point. If we see that it has been below and has found an abnormal amount of gold for that pumping we can be sure that something was wrong—some gold has run into the pipe. In this hole, the bit is never below the casing except after the casing has been driven 0.5 ft. into bedrock. The reason for this is, that the machine-man is under orders to cease drilling while there is yet 0.3 ft. of core in the casing. There is only one reason for the bit to be below the casing; the ground is so tight or so coarse that the casing will not drive unless the ground is loosened below the casing. And in no case where the ground is drilled below the casing should the material be pumped before the casing is again driven deeper. This core is to act as a plug to prevent any material from outside the limits of the shoe from running into the casing. Its length will depend on the ease with which the ground runs. In sand, a larger core is required than in gravel. The proper length of core to leave can be determined only in the ground itself and will vary not only for different holes but also for different depths in the same hole. This must be regulated by the man in charge.

The length of core left in the pipe before driving should always be such that only the right volume of material will be forced into the casing. By the right amount is meant that amount which corresponds to the volume of a cylinder one foot long and 7½ in. diameter at the base. This amount, when forced into the pipe, should be somewhat longer than one foot, as it has been reduced at the base to a circle only six inches in diameter. This cylinder, to have the same volume as the one of larger cross-section, not allowing for the expansion due to the loosening of the gravel, would be 18.7 in. long, or 1.555 ft. But only in those cases where the ground is classed as "very firm" does the length of core approach this value. (See column "B" for the following feet: 15-19 and 27-30).

In column "D" is placed the core after pumping, and its importance in one way has been shown. Its further importance will be indicated later. In column "E" is placed the depth of the hole. It is obtained by subtracting the core in "D" from the depth of the casing as seen in "A." It is the least important of all, and is only retained as a convenience.

TABLE I

In column "F" are recorded the estimated amounts of gold, which are classified as explained above and are reported by weight, not by number of colors. For example: 3^7 means that seven milligrams of No. 3 gold was found in that pan; 2^9 means nine milligrams of No. 2 gold, and 1^{18} indicates 18 milligrams of the largest size were in the pan. This estimating of weights tells a great deal more than the mere number of colors. It is remarkable how expert a man will become in estimating the gold, especially in the smaller sizes. In looking at the log-book one can immediately tell how the ground is running and how many cents it will pan per cubic yard. Each milligram of gold per running foot

TABLE 1

A Ft.	B Ft.	C Ft.	D Ft.	E Ft.	F—Gold Contents	G In.	H Hr. Min.	I	J	Remarks
4.7	2.5	0.3	0.3	4.4	— — —	1.00	7-56	Loose	Soil.	M. Blk. S.
6	2.1	0.4	0.3	5.7	— — 3^{1}	1.00	8-04	"	S. F. G.	"
7	1.5	0.3	0.3	6.7	— — 3^{7}	0.75	8-24	"	"	"
8	1.5	0.5	0.5	7.5	— — 3^{2}	0.75	8-32	"	"	L. Blk. S.
9	1.7	0.3	0.3	8.8	— — 3^{1}	0.75	8-42	"	S. Md. G.	"
10	1.8	0.3	0.3	9.7	— — 3^{3}	0.75	8-50	"	"	Sm. Blk. S.
11	1.8	0.5	0.7	10.3	— — 3^{6}	1.00	9-03	Firm.	S. C. G.	"
12	2.9	0.7	1.2	10.8	— — 3^{4}	1.50	9-16	"	Sand.	M. Blk. S.
13	2.7		0.5		— —					
14	2.8	0.6	0.5	13.5	— — 3^{5}	2.00	9-47	"	S. C. G.	Sm. Blk. S.
5	1.8	0.4	0.4	14.6	— — 3^{1}	1.00	9-59	"	"	"
16	1.8	0.3	0.3	15.7	— 2^{2} 3^{3}	1.00	0-20	V. Firm.	C. G.	Sm. Clay.
17	1.8	0.3	0.3	16.7	— 2^{6} 3^{5}	1.00	10-42	"	"	Sm. Gs.
18	1.8	0.4	0.4	17.6	— — 3^{4}	1.00	11-12	"	"	"
19	1.8	0.3	0.4	18.6	— — 3^{7}	1.50	11-20	Firm.	S. C. G.	Sm. Blk. S.
0	2.5	0.6	0.6	19.4	— — 3^{7}	1.50	11-31	"	Sand.	"
21	2.0	0.5	1.0	20.0	— — 3^{4}	1.50	11-43	"	S. F. G.	"
2	2.2	0.7	0.6	21.4	— — 3^{3}	1.75	11-56	"	"	"
3	1.8	0.4	0.3	22.7	— — 3^{7}	1.00	12-40	Loose	F. G.	"
4	1.5	0.3	0.2	23.8	— — 3^{7}	0.75	1-05	"	"	"
25	1.6	0.3	0.3	24.7	— — 3^{1}	0.75	1-17	Firm.	Md. G.	"
6	1.7	0.3	0.3	25.7	— — 3^{3}	0.75	1-23	"	C. G.	Garnets.
27	1.8	0.3	0.3	26.7	— 2^{2} 3^{6}	0.80	1-45	V. Firm.	"	"
8	1.8	0.3	0.3	27.7	— 2^{3} 3^{6}	0.75	2-03	"	"	"
9	1.8	0.4	0.4	28.6	— — 3^{2}	0.75	2-33	"	"	"
0	1.9	0.3	0.5	29.5	— — 3^{6}	1.50	2-51	"	"	"
31	2.3	0.5	0.7	30.3	— — 3^{2}	1.50	2-57	Loose.	S. F. G.	Sm. Blk. S.
2	2.2	0.6	0.6	31.4	— — 3^{7}	1.50	3-07	Firm.	Sand.	"
3	1.9	0.5	0.4	32.6	— — 3^{7}	1.25	3-14	Loose.	S. F. G.	"
4	1.7	0.3	0.3	33.7	— 2^{6} 3^{9}	1.00	3-26	Firm.	C. G.	Garnets.
35	1.9	0.4	0.4	34.6	1^{6} 2^{9} 3^{12}	1.00	4-10	V. Firm.	C. G. & Bedrock	
10—26—'04										
6	2.3	0.5	0.5	35.5	1^{27} 2^{18} 3^{18}	1.00	7-25	"	Bedrock	
		—1.0	—1.0	37.0	— 2^{6} 3^{8}	0.50	7-32	Soft	"	
		—2.0	—2.0	38.0	— — 3^{4}	0.25	7-40	"	"	
		—3.0	—3.0	39.0	— — —		8-00	"	"	

means that that foot corresponds to ground of a value of 6c. per cu. yd. For example, in the 15th foot the panner recorded two milligrams of gold and one immediately knows that that corresponds to 12-cent ground; in the 17th foot there were 11 milligrams or that ground was 66-cent ground.

This value is obtained thus: Each running foot means 0.01 cu. yd. Therefore, if we get one milligram from 0.01 cu. yd., from a yard we should get 100 mg.; and as a gram of gold (as ordinarily found in the gravels of the San Joaquin and Sacramento Valleys) is worth about 60c., 100 mg. is worth six cents. This method of recording weights emphasizes the occurrence of pay-streaks much better than that of recording colors only.

In column "G" is recorded the amount of material obtained by pumping, as measured in the pan. It is a rough check on the volume extracted and its use is shown later. In column "H" is given the time of each pumping. It helps to afford an idea of the stiffness of the ground by a measure of the time it takes to drive, drill, and pump a foot. In column "I" is recorded the character of the ground in regard to its tightness. Ground is classed as "loose", "firm" and "very firm." "Loose" ground is that which can be drilled and pumped at the rate of one foot every 5 to 9 minutes; "firm" ground would be such as required from 10 to 14 min. for the same operations, and "very firm" that which required a longer time. In column "J" is recorded the formation as passed through. This is classified under the following heads: Clay, sand, cemented material, fine gravel, medium gravel, coarse gravel and large boulders.

Under the head of remarks, almost anything important or unusual is recorded. This would include the amount of

TABLE II

Hole No. 29. C$_2$ W. J. Smith's Tract.

Commenced hole	7 a. m. 10-25-1904	
Finished drilling	8 a. m. 10-26-1904	
Finished pulling	9:47 a. m. 10-26-1904	
Finished hole	10:05 a. m. 10-26-1904	
Depth of hole	39 ft.	
Depth to bedrock	35.5 ft.	
Character of bedrock	Soft white tuff, much slime	
Depth of soil	4 ft.	
Water level from surface	14 ft.	
Total colors	1^{33}, 2^{52}, 3^{119}, 204	
Gold in box	— — 3^{4}	
Gold in tails	— — 3^{4}	
Pay-streaks	8-18, 21, 22, 25-31, 34-38	
Delays—cause and length		
Length of pipe—No. 1	6.5	....
2	6.1	12.6
3	4.8	17.4
4	5.3	22.7
5	5.7	28.4
6	5.6	34.0
7	5.1	39.1
Sand	5, 12, 20, 32	
Clay	— — — —	
Cement	— — — —	
Fine gravel	6-8, 19, 21-24, 31, 33	
Medium gravel	9-10, 25	
Coarse gravel	11, 13, 18, 26-30, 34, 35	
Loose gravel	5-10, 23-24, 31, 33	
Firm gravel	11-15, 19-22, 25-26, 32, 34	
Very firm gravel	16-18, 27-30, 35	

Remarks.—A great deal of sand occurs mixed with gravels and in places there is enough water to cause the ground to run.

Calculated volume	0.38 cu. yd.
Volume by cores	0.45 cu. yd.

black sand, the occurrence of amalgam, etc. Here also would be mentioned anything peculiar about the gold, such as its lightness, both in weight and color, and the occurrence of flour gold. The delays, with their causes and lengths, would also be stated here.

TABLE II

Table II is the summary which follows each hole. It is self-explanatory. The water-level is measured during, or after, the casing is pulled. If there is no danger of the ground caving, this is left till but one joint remains in the ground. If there is any danger of the ground caving, the measuring is done after each joint is pulled.

I think I have shown how the drill and its record can be made to answer all of the first four questions, with the exception of the value of the land and the contour of the bedrock. These are obtained by the use of the records in the log-book with those obtained in making the surveys. By running a line of levels over the holes, and using the depth to bedrock as found in the log-book, the contours of the bedrock can be estimated.

CALCULATING VALUES

In figuring the value of the ground, one must be certain of the volume of material from which each sample came. It is the lack of this knowledge which has caused most of the errors made in valuing ground by the method of boring holes. The volume of a hole excavated as described is not that of the inside of the casing, but is the volume of a cylinder whose length is the distance from the surface to bedrock, or as deep as pay goes, and the area of whose base is

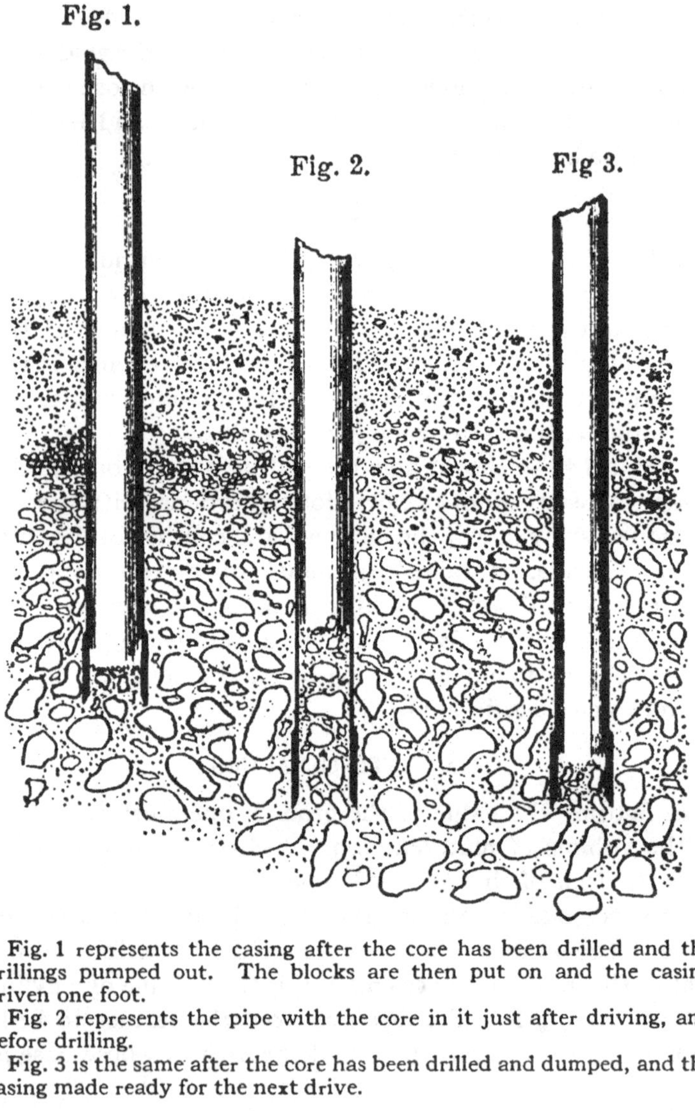

Fig. 1 represents the casing after the core has been drilled and the drillings pumped out. The blocks are then put on and the casing driven one foot.

Fig. 2 represents the pipe with the core in it just after driving, and before drilling.

Fig. 3 is the same after the core has been drilled and dumped, and the casing made ready for the next drive.

the area of a circle 7.5 in. diameter; that is, the diameter of the shoe. If no ground has run into the pipe, we can use what is called Radford's "Keystone" factor; that is, 0.27. This is used in the following way: Multiply the depth in feet by 0.27 to get the volume in cubic feet. To reduce to cubic yards, divide by 27. To simplify matters, divide the depth of the hole by 100 to get the volume in cubic yards. This factor was obtained by experiment in one case, and apparently does not take into consideration the possibility of any other material, not belonging to that cylinder, running in. This ground that has run into the pipe carries gold and must be taken into consideration in figuring the volume from which the sample came.*

The weight of the gold from the hole in Table I was 184 mg., or approximately 11.04c.; the volume for 38 ft., according to Radford's "Keystone" factor, was 0.38 cu. yd.; therefore, the value of a cubic yard is 29.05 cents.

*[No attempt is made to weigh or measure the volume of material actually excavated from the bore hole. Such a method would introduce many elements which would usually make the results wholly unreliable. As a matter of fact some good engineers have gone wrong at this point by comparing the value of the gold obtained from the test with the volume of the material delivered in the sluice box by the sand pump. The reason why such a method would be unreliable is, that when the material encountered by the drill, drive shoe or sand pump has been disturbed, mixed with water and allowed to settle, it will not give the same cubic volume which it originally occupied. There would be a change in the proportion of water contained and the volume would also be changed by a rearranging of the particles, especially if the drillings consisted in part of clay and in part of sand. A bucket full of fine clay may be mixed with a bushel of coarse sand and gravel and the whole put in a bushel measure. In certain soils, a post hole may be bored, a large-sized post inserted and all the borings tamped in about it without more than filling the hole. In making these alluvial tests the strata may change from dry to wet or the material from clay to sand, or from fine to coarse sand many times between surface and bedrock, so that the volume tests or volume method of valuing the ground is wholly unreliable. The fact that the drillings actually taken from 100 feet of test hole measure more or measure less than that which is represented by the core for that depth does not at all indicate whether the drill has excavated more or less than the area of the drive shoe.

From all the above, it will be clearly apparent that in making the tests everything depends upon having tools and appliances that will excavate the exact volume represented by the drive shoe from surface to bedrock, or at least while passing through that part of the distance in which the gold is to be found. Making hole to bedrock is not making a test. If, in making the hole, some of the material is driven off into the side of the bore hole the test will show too little gold. If, on the other hand, the surrounding material with its gold is drawn into the bore hole the amount recovered will be too great.—Keystone Driller Company.]

In looking at the cores in column "B", it is seen that they differ for the different formations drilled. For example, compare the cores in columns "B", "C" and "D" for the following feet with those of any of the others: 11, 12, 13, 19, 20, 21, 22, 30, 31 and 32. In "B" they are larger than ordinary, and in places in "D" we see that the core after pumping is greater than before. There is only one conclusion; the ground is running into the pipe and here we are getting more than is represented by a cylinder 38 ft. long and 7.5 in. diameter. This must increase the volume, and it certainly increases the gold obtained, as can be seen by looking at column "F." This increase in volume must be taken into account or else the increase in the amount of the gold will give a false return.

By a careful use of the core-lengths recorded and the volumes given in "G", the volume of the incoming ground is obtained thus: Opposite the 11th foot, the core after pumping is 0.7 ft. and before it was only 0.5 ft.; clearly a running-in of 0.2 ft. Then, too, in column "G" we get one pan of material where we have only been getting 0.75 pan before. That shows that a little more was pumped than was to be expected by the core after driving.

Again in the 12th foot we have a running-in of 0.5 ft. after pumping, and to begin with we had 2.9 ft. core where we did not look for more than 2.2 ft.; therefore, for this foot alone, we have a running-in of 1.2 ft. By going down all of the cores we see that there has been a running-in of about 5 ft. Then by looking at the pans we see a greater amount of material than was to be expected from the cores. For this hole, this amounts to about another two feet; that makes an extra 7 ft., or the hole corresponds to a depth of 45 ft. and the volume would be 0.45 cu. yd. Using this as our basis of valuation of the ground, we get an average of 24.5c. per yard. This is about 15 per cent lower than the yield we found above and I think is much nearer the true value. Such a difference would destroy the valuation of a large deposit.

Of course, all holes do not contain running ground and where they do not, there is not the need of making these modifications. For this class of ground, this form of log-book seems to record the facts better than a simpler type. As one cannot tell when running ground is to be encountered, this form of notes should be used all the time.*

FINAL CALCULATIONS

The average value of the ground for the whole area is obtained in the following way: Multiply the average yield of the ground for each hole by its depth in feet; add these products and divide the sum by the sum of the depths. The volume is obtained as soon as the area is known and the average depth over that area. The gross value of the

*The "Keystone Rule" for computing values is as follows:

Multiply the value of the gold (in cents) by 100 and divide by the number of feet drilled—the result is the value of gold per cubic yard.

After experimenting with larger and smaller sizes we have selected the 6" test tube as the smallest which will give reliable results. This tube has a (nominal) inside diameter of 6" and is shod at its lower end with a tempered steel shoe, the cutting edge of which when new and sharp is 7½ inches in diameter. It is the size of the shoe and not the bore of the pipe which should be used as basis for computing values. The area of this 7½" shoe at its cutting edge is about 3-10 of a square foot. Hence for every foot in depth there will have been excavated 3-10 of a cubic foot, or (since there are 27 cu. ft. in a cubic yard) 3-270 or 1-90 of a cubic yard. In operation, however, the cutting edge of the shoe may and does become more or less battered or bruised (by contact with boulders, etc.), so that as practice and observation have shown, it is more nearly correct to subtract from this amount about 1-10 which would make the amount excavated from each part drilled just about 1-100 of a cubic yard, or one cubic yard for each 100 ft. of test hole.

An Example and Answer by the Keystone Rule.

Suppose the test tube has been put down 38 ft. Total amount recovered from drillings say 12c. Multiply the number of feet (38) by the Keystone Factor (1-100) and we have as the amount of the material excavated 38-100 of a cubic yard. Then if 38-100 of a cubic yard produces 12c worth of gold, the amount obtainable from the whole cubic yard would be 100-38 of 12c or 31 6-10c (nearly) per cubic yard.

Moving with derrick up

deposit is then found by multiplying the value in cents per yard by the total number of yards. The net value is obtained by subtracting from the gross value the cost of the land, the cost of working, and the loss in tailing. These will differ under different conditions and in different parts of the country.

A machine will make from 12 to 40 ft. per day; it is impossible to give an average, as it varies in different parts of the State.

VALUE OF TESTS*

Drilling followed by dredging, as at Oroville, has shown that only about 75% of the yield shown by drilling has been recovered. To test the value of these tests one acre was drilled with 23 holes and the dredging returns produced bullion worth 95% of that shown by the drilling results. In another case, a boat dredged only about 35% of the gold as shown by the drilling, but in this case there were only two holes per acre. It is plainly seen that the more numerous the holes, the more accurate is the sampling. With care and a careful superintendence of the drilling, there is no reason why these results should not be fairly accurate; they can always be checked by the sinking of shafts preparatory to the commencement of actual dredging.

(End of Mr. Stines' article.)

*As to prospecting ground with drills, it is said that the Oroville experience has shown a yield by dredging from 70 to 85 per cent of that given by drilling tests. It is just such statements as this that are useless, although apparently businesslike. The value of a series of drill-holes as indicating the richness of a tract of gravel cannot be gauged by any empirical formula. It is as absurd as the practice of timid engineers who cut their estimates of ore in two or deduct a certain fixed percentage from their calculations, so as to be safe. In the case of drilling before dredging, the result is reliable according to the number of holes, the distribution of them, the care taken in the work and, above everything, the personal factor. Such work carefully done and checked at each stage by a man of experience and integrity needs no big discounting, while that accomplished by an unreliable driller or a careless novice is worse than worthless. No fixed percentage covers the varying conditions surrounding an engineering operation.

In the early days of dredging it was the custom to test 200 acres with 10 drill-holes; nowadays one hole to two acres is not uncommon. When results indicate that the gravel is of varying depth, that there are irregular channels or that the ground is spotty, it is not unusual to put down as many as one to three holes per acre, especially where no adjoining workings exist, such as throw light on any anomalous results from the drilling. The cost is a limiting factor, for each hole costs $60.00 to $100.00 in ground of any considerable depth.

It is generally assumed that the results given by shaft-sinking are more reliable than those from drill-holes; as a rule this is true, but the comparison too often smacks of the old idea that a mill-run is more trustworthy than the sampling of ore in a mine. It depends upon how it is done, with this proviso, that the more men needed to carry out an operation of sampling, the greater the opportunity for error, intentional or unintentional. A shaft gives better opportunity for examining the nature of the successive layers of gravel and other conditions that bear upon the subsequent working of the ground. Careful measurements are imperative; sometimes it is not practicable to hold the same diameter of shaft all the way down, and in running ground one has to resort to timbering; these factors affect the cross-sectional area and must be carefully noted in any calculations. Moreover, instances are known where shaft results have been seriously vitiated by the fact that the particles of gold have fallen to the bottom, with the water, so that the upper layers of gravel appeared worthless, while that immediately above bedrock was excessively enriched. Great caution, bred of experience, is required to make an accurate test.—Editorial from Mining and Scientific Press for February 3, 1906.

ACCURACY OF THE TESTS

This has been proved by experiment in many ways, the most conclusive of which was made at Oroville, California—where there are now (1907) about 40 large dredges at work upon large areas of ground—all assayed in advance with our machines. A test tube was first sunk in the usual way with one of our machines. The contents of the test hole was computed, and the amount of gold washed out of it weighed carefully. Then to prove the accuracy of the results thus obtained, and without moving the test tube, a three-foot shaft was sunk around it, the sand and gravel washed, the gold taken out of it and weighed. The results of the two tests were then carefully compared and it was found that they tallied to within three to five per cent—the tube test giving from three to five per cent less gold than the shaft excavated by hand. The difference was easily accounted for by the uneven surface of the hand-made shaft, it being impossible to excavate by hand to the exact size. The tube test as made with our machine was therefore considered the more accurate.

In other cases, where the gold was confidently believed to exist and none was found with the test tube driven, small quantities of gold dust were weighed and dropped into the tube when the hole was about 25 feet deep. The gold dust was then thoroughly mixed by cutting up several feet of material with the drilling tool and driving the pipe that distance, but not to bedrock. The cut-up materials were then recovered with the vacuum sand pump and panned. The gold recovered was dried and weighed and corresponded within a very small percentage to the respective amounts put in. In this way, in one particular instance (near Marinsk, Siberia), it was proved that the hand-made shafts previously excavated by the natives, and which had assayed "high values", had been most cleverly "salted" by the owners for selling purposes. The discovery of the trick spoiled the sale and resulted in the cancellation of the order for about a quarter of a million dollars' worth of dredging machinery, but we had the sincere thanks of the purchasers of our machines—as it saved them wasting a vast amount of money. Our tests proved the ground wholly barren.—Keystone Driller Company.

PROSPECTING FOR COPPER WITH CHURN DRILLS

Written for The Mining and Scientific Press (Issue of Dec. 29, 1906) by F. S. Pheby

Believing that the great horizontal ore-bodies of the Ely district could be more cheaply and expeditiously tested by the use of the drill rather than through the usual tedious and expensive method of shaft-sinking, the Ely Central Copper Company purchased two drills, manufactured by the Keystone Driller Company, of Beaver Falls, Pennsylvania. These drills were hauled from Cherry Creek to the mines, a distance of 60 miles.

These machines are type No. 3 and weigh 12,000 pounds each. A complete outfit of tools and equipment for recovering any parts of the drills lost in boring the holes were included in the order.

The first drill was installed on the property in August, and we had a hole completed to a depth of 308 feet within 23 days, although but one shift was engaged in this work. It is far more economical in wood and water to keep the drills running continuously with three shifts. More than three times the sinking can be done in the same number of days, as steaming up in the morning generally comes out of the one shift. The drill averaged 13.3 feet per shift for the 23 shifts, but if shut-downs were deducted, the drill averaged 19.6 feet per shift of actual operation.

We have found in all our work that casing is necessary in every hole. At times we have sunk 200 feet without casing, but to go deeper, casing was necessary, which meant that the

Mr. Odell and machine testing for gold in Idaho

hole had to be reamed out at an expense of more than the original cost of the hole for the 200 feet. I caution anyone contemplating the purchase of a drill, to provide 7⅝ casing for one-third the depth contemplated, 5⅝ casing for two-thirds the depth, and 4¼ casing for full depth. This we had not done, and probably 25% has been added to the cost of each hole, working without sufficient casing. Trouble may not be encountered in several hundred feet, but a soft stratum of ten feet will require the casing for the entire hole.

We soon learned that the wear and tear on the drilling cable was severe. Under ordinary circumstances, each rope is safe for 1500 feet of drilling. When we appreciated that our ropes were deteriorating, an order was placed at once. By October 1st the cable had not arrived, and we were compelled to order by express from Ohio a coil of rope weighing 1850 pounds. With each drill, two ropes of the same length as the depth of the hole contemplated should be ordered.

Water is required both for the boiler and for diluting the drillings so that the sand pump may bail them. The greater quantity is required for the boiler. This item of expense is local with each hole. It is advisable to haul water in a good wagon-tank, as the drills may be moved many times, and the cost of pipe might be greater than the cost of hauling. A drill-man and helper are required on each shift, the former receiving $4.00 for eight hours' work and the latter $3.25. The items of fuel and water are wholly relative, but will be given in our particular case, and may be taken as an average for this district. Figures are for 1906.

The following is the tabulated costs for a certain hole which we have taken as an average:

Shifts of sinking (eight hours each)	23 days	
Depth of hole	308 feet	
One drill-man, wages		$ 82.69
One helper and assistants while handling casing		81.08
¾ cord of wood each shift, at $4.50 per cord		77.51
12 bbl. water at $6.50 for hauling, 8 days		52.00
12 bbl. per day (when running two rigs) 15 days		48.75
Coal and oil		7.60
Miscellaneous charges		12.20
Superintendence		50.00
Cost of 308-foot hole		$411.83
Cost per foot		$ 1.33

Some water was encountered in this hole, and I may say roughly that a two-compartment shaft for the same depth with equipment, would have cost about $12,000, or $40.00 per foot.

The question has often been raised concerning the character and accuracy of the sample obtained from this work. With small rich veins there may be an objection to the use of drills, but in orebodies like those found at Ely, I believe as good a sample can only be taken with great care. Most of the sludge or drillings will pass a 20-mesh screen, and a good method is to provide a large box with a capacity equal to several screw-lengths. I might say that a screw is three feet long, and when fed out, the clamps are changed on the rope, and the hole bailed, or sand-pumped. By settling and decanting the water, the entire product of the hole may be saved and sampled. In ordinary practice, it is sufficient to dip a sample from the box and pour the same into a box partitioned off in compartments about the size of a common brick. This sample, when dried off in the sun, is compact enough to be sampled by chipping, and can be shipped and carried about without breaking.

In a hole carried down without casing, there is danger of knocking down particles from the upper portions of the hole. This will vitiate the sample, but the harm done is more theoretical than actual. As is well known, the orebodies of Ely are the impregnation of a great stockwork of porphyry, and an orebody of much value must be a hundred or more feet thick. While actually in ore, the sampling must be done with extreme care, but often the hole, all or in part, may be in barren country rock, and only a knowledge of the formation penetrated is desired.

The best test, where not in ore, is obtained by panning the sample. This concentrates the coarse particles, which are clear to the eye if a magnifying glass of low power is used. In oxidized formations some doubt often arises as to whether the drillings are composed of porphyry or limestone. A small bottle of dilute hydrochloric acid will soon settle this question with entire satisfaction. We have made a practice of saving a sample of each formation penetrated, in four-ounce bottles. This gives a clear picture of the hole, and is valuable for future reference.

Our drills are provided with traction gears, and may be moved at will over the roughest ground. This saves the expense and trouble of procuring teams. The winter weather is severe for outside work, and sectional houses, so made as to be readily knocked down and moved, have been constructed to enclose the drills, and within these houses the men work in comfort.

A drill of this size is light for holes of greater depth than 500 feet. The size recommended is a No. 5, good for 1200 feet, and the cost of the same with traction gear is about $2500 laid down at Ely. Should five holes of 500 feet each be sunk, the cost of equipment per hole will be $500, about the price of a good whim outfit. The cost of $1.50 per foot is so much cheaper than any shaft work as not to be comparable. Water does not retard the work of drilling; in fact, it is a benefit.

I can heartily recommend the use of these drills for prospecting in this region. Apart from the cost per foot, the time in which a good working knowledge of the ground can be obtained, is a great factor. The sample is quite as good as a core, and most ground can be penetrated at 25% of the cost incurred with a diamond drill; while the expense of equipment will not exceed 30% that of the diamond-drilling outfit.

BLASTING TIGHT PLACERS BEFORE DREDGING

By Oliver B. Finn

From the Engineering and Mining Journal of July 7, 1904

Gold dredging has made such rapid strides of late and there is such a widespread interest in this branch of mining that I venture to contribute a detailed account of the way in which a Keystone driller was used by me in California to loosen, by blasting, a very tight gravel deposit, preparatory to dredging. There is a great difference between "cemented gravel" and "tight gravel." A truly cemented gravel is not a dredging proposition, while the tightest possible gravel, where there is no cement, can be made easy working by the following method of blasting:

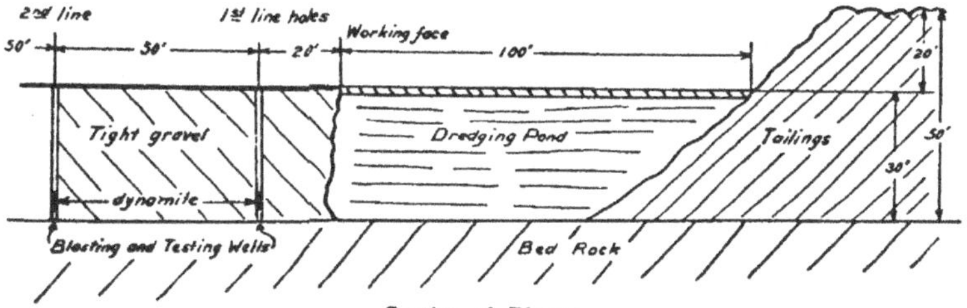

Section of Placer.

The deposits in question were practically a solid mass of cobbles, the voids being filled with a heavy sand, forming a mass so tight that every conceivable method of manipulating the dredge resulted in constant breakages and a failure to dig sufficient gravel to yield a profit. The dredge was of

On the Yukon

the elevator type, much too light in construction for unblasted tight ground. The results of an eight months' run showed the average load of the buckets to be only one-fourth of their capacity, while much of the time many of them ran empty, and 50 per cent of the working time was lost in shut-downs for repairs, thus reducing the work done by the dredge to one-eighth of its capacity. The repairs had cost thousands of dollars; and although the ground contained good dredging values, the undertaking was a total failure, until finally blasting was adopted. It was proposed to blast this ground, using a 6-in. churn drill to sink wells (in the bank ahead of the dredge) in which to place the explosives. As the procedure was entirely new, a number of months were wasted before the owners of the dredge would supply a drill, but being prevailed upon to do so, upon arrival of it, a line of holes was driven to bedrock about 20 ft. back from the working face; the holes 50 ft. apart, and each succeeding line of them 50 ft. farther back. The holes in each line "staggered" with those of the next. Each hole was blasted as soon as completed. Black powder was found useless, since, owing to the tightness of the ground, the whole force of the explosion was spent in blowing out the tamping and water. Various sizes and strengths of explosives resulted in the selection of 30 lbs. of 80 per cent nitroglycerine dynamite. This was put up to order at the factory in tin cans 3 ft. long and 4.5 in. in diameter, with a wire bale fastened to the inside of the can so as to make a smooth cartridge readily inserted into the well through the inside of the casing. Before inserting the explosives, the well casing was drawn up about 4 ft. from bedrock; the casing could not be entirely drawn before loading the hole, as the hole

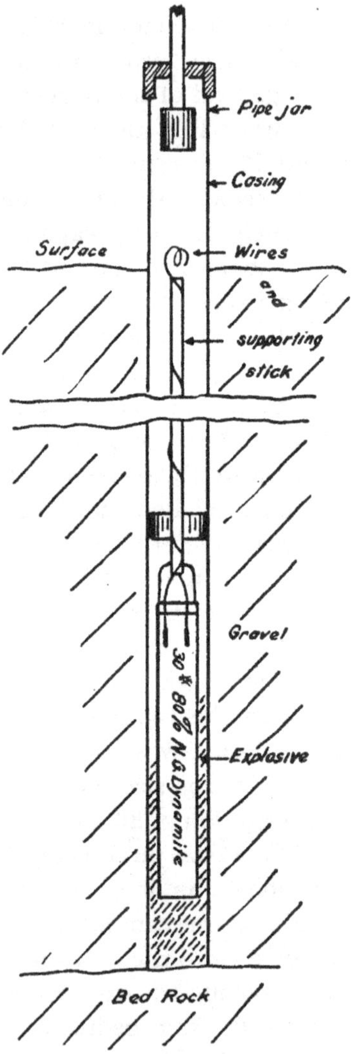

Method of Blasting.

would be partially filled by caving in from the sides. Electric exploders were used, two to each charge; the second one as a precaution against a missed shot.

As the "pipe jar," which is the best device for pulling up the casing, practically closed the top of the casing, the electric wires had to be kept within reach during the drawing of the balance of the casing by fastening them to a stick one inch square, which reached from the cartridge bale to the surface of the ground when the cartridge was in place, care being taken to have sufficient casing above ground to prevent the "pipe jar" from striking the upper end of the stick. The stick also served the purpose of supporting the charge, when lowering it into the well and preventing any strain upon the electric wires. When the charge had been carefully lowered into place, the "pipe jar" was replaced and all casing withdrawn. The hole was then filled with sand, and the drill moved to the place for the next hole. The wires were connected with the dynamo upon the dredge, and the charge exploded.

The explosion did not displace any of the bank, but the shock and vibration were such as to loosen the cobbles without disturbing their original relative position, excepting, of course, those close to the shot. Each explosion was effective far beyond its allotted area, so that the spacing of the holes could be gradually increased to as much as 100 ft. apart.

By saving the gold from the drillings of each hole, this method served the double purpose of blasting and testing.

This preparation of the ground resulted in the buckets running full to overflowing continuously, and there was no occasion for shut-downs, except for clean-up of gold and oil-

Prospecting in Russia

ing the machinery; notwithstanding, the machine was practically a wreck from previous hard usage. In other words, the capacity or output was increased eightfold, and the expense for repairs reduced to that of normal wear and tear. In this way many total failures could be turned into successes, and the necessity avoided of either giving up the entire project or supplanting a dredge otherwise too light for the work with a heavier and more expensive one.

The average cost of the blasting was not over 0.5c. per cubic yard. Where ground is at all tight and the buckets do not easily fill, I would advocate blasting of ground, even for the most powerful machine that could be built; since it is evident that a very small outlay in this direction would be saved many times over in power expended on the bucket line and in lengthening the life of all wearing parts, to say nothing of the increased capacity of the dredge.

[End of Mr. Finn's article]

PART III

MINERAL PROSPECTING MACHINERY

BY

R. M. DOWNIE

General Manager, Chief of the Engineering Department for Forty Years and Founder of the Keystone Driller Company in April, 1882.

MINERAL PROSPECTING MACHINERY

By R. M. Downie

General Manager, Chief of the Engineering Department for Forty Years and Founder of the Keystone Driller Company in April, 1882

CORE DRILLS

We have it upon the authority of its Maker that "the earth is full of riches." Experience has proved this statement true, and the machinery herein described has been planned and perfected for the purpose of exploring for it.

New store bins of wealth are being constantly discovered in all parts of the world and there seems to be really no limit to them. No section or country has, or ever can have, a monopoly of mineral wealth. The reason some portions seem to have more than others is frequently because the people of those localities have the enterprise to investigate and develop what they have.

The older rocks and their conglomerates carry the precious metals; the newer contain zinc, lead and iron; and those of more recent formation contain oil, asphaltum, gas, coal, fire-clay, etc. The drill is the key that unlocks all these treasures—coffers whose robbing wrongs no one.

Prospecting Drills in general are made to operate upon two well known general principles. The older and best known of these is the Percussion Drill, which penetrates the earth by a succession of blows with a cutting chisel. This is the cheapest known method of drilling, and generally the quickest and surest particularly for deep borings. Such a drill cuts up the material in its path; but with the following exceptions, does not take out a core or solid section of the strata. The Keystone Percussion Core Drill can be used on any of our machines without any change in the operating machinery. This combination makes an ideal outfit for prospecting coal measures, etc. The regular solid percussion drill bit can be used until the mineralized stratum is reached. At this point, it and the solid drill stem used with it can be taken off and, in their place, the Core Bit substituted and operated without any other change. This Percussion Core Bit is made in several forms to suit different kinds of rock. The joint (or screwed connection by which it is attached to the Rope Socket or Jars), is made to fit any drilling tools of our manufacture and can, if the size is furnished to us, be made to fit the joints of any other make of machine. It and its operation are fully described in a catalog, which will be mailed upon application.

The other general form of prospecting drill is the "Revolving Core Drill", which operates by being revolved like a wood auger. It is made in three different forms, all of which act on the same general principle. The first and most common form of it is the diamond drill. It is made by inserting sharp cornered diamonds in the thickened end of a tube. These diamonds act somewhat like the teeth of a saw, and in their revolution, cut or scrape a circular channel, leaving in the center a "core."

A second form of the revolving core drill is made of steel or other metal (to escape the expense of the diamonds), and consists of a series of saw-tooth points. It operates by scraping out a circular channel about a center or core. It works fairly well in formations which are soft and devoid of grit; but in any material hard enough to make a grindstone the points of the teeth are immediately ground off and operations cease until the teeth are renewed.

A third form, designed to avoid the expense of diamonds and the wearing of the steel teeth, operates to cut the channel by revolving the tube upon a layer of chilled steel shot introduced constantly in the bottom of the bore-hole. But, except where the material is devoid of crevices, this latter form has met with but little success even in getting the hole down. Other difficulties are noted below.

With all these three forms of revolving core drill, and when the rock is suitable and seamless, the core remains within the revolving tube still attached to the bottom of the bore-hole until it is from one to six feet long. If this core is not destroyed in the operation, a "core catcher" grips it near the bottom, breaks it loose from its anchorage and brings it to the surface for examination. During the process of drilling with the revolving core drill, in order to wash away the cuttings, prevent gumming, and in order also to keep the diamonds, teeth or shot, as the case may be, from being heated, burned or torn from their settings, a stream of water is constantly forced down the hollow drill stem by a pump. The cuttings are very finely pulverized and are generally washed up the outside of the drill or flow away beyond recovery through openings in the rock. Thus, the entire result obtained by such a drill is summed up in the core.

Allow us to say at the outset that we do not manufacture any form of revolving core drill. We refer to them here merely for the purpose of pointing out the peculiar conditions under which favorable results may be expected from them. The only forms of core drill which we make are operated by means of a cable with short up-and-down blows and are usable interchangeably with the regular percussion drills described at length in this catalog. Neither can these revolving core drills be operated by our machine without the addition of revolving mechanism and pumps for forcing water down the hollow drill rods. The foregoing remarks are introduced solely for the purpose of pointing out the difference and explaining exactly what we do produce. This will save written correspondence and enable intending prospectors to intelligently decide what sort of machine and appliances are best suited for the work to be accomplished.

Where the natural conditions exist which allow the core drill to be operated and which permit the core to be certainly recovered, there is nothing more accurate or satisfactory. It shows the texture, stratification, laminations and dip of the rock in undisturbed section; but there are conditions under which it is wholly useless in any form.

It is a good and sound business policy to sell a customer exactly what he needs and machinery and tools which will perform with certainty and exactness the work desired. This can be accomplished only by making the customer fully acquainted beforehand with all the conditions of success and by guarding against disappointment by pointing out also what can not be done or expected. Hence, we pledge our customers that we will not, under any circumstances, knowingly sell them that which will not meet their requirements in the best possible way. Therefore, we ask prospective customers to inform us fully and concisely what they wish to do—and if, as sometimes happens, we do not make the machinery and tools best suited to their requirements—we will say so, frankly, and refer them to the makers of that which we believe will do it. If a positive order comes to us for goods listed, without any explanation, we take it for granted that the customer knows his business, and act accordingly.

But we wish it understood, (1) that we have every facility for correspondence, and (2) those in charge of our several departments are men who have had in the field a wide practical experience in the operation of the machines. We can therefore advise prospective purchasers with an "assurance born of a knowledge of having done it."

It goes without saying that information imparted to us for the purpose of adjusting the machinery to the work required is considered confidential and in no case divulged.

THINGS WHICH CANNOT BE DONE WITH A REVOLVING CORE DRILL

Before proceeding to a description of our machinery, its various uses and plans of operation, let us briefly note some things which cannot be done successfully with a core drill.

1. A hole cannot be made through sand, gravel and boulders—because the loose formations will immediately close up the bore-hole upon the withdrawal of the drilling tool.

2. With a core drill a casing cannot be inserted to keep out the loose formations because it provides no means for driving the pipe or breaking up the boulders ahead of it. Of course no samples or core can be taken out of such formations with a core drill, except what may be washed up the outside of the tool; and even the material which might be thus recovered leaves the operator to guess whether it comes from the bottom of the bore-hole or has been dislodged from the sides of the well by the upward rush of the water.

3. A revolving core drill will not with any precision or certainty, go through broken, seamy rock; and it is wholly inoperative among boulders. The hollow shaft of the drill does not fit the bore-hole closely, and hence it may be easily deflected by a hard surface, like a gimlet diverted by a nail or knot; or it may seek to follow a soft lead or fissure. Then there is nothing in the principle of its operation to draw the drill back to a straight line if such deflection once takes place; and hence a slight departure from a right line will be cumulative as depth is gained, until the hole is presently too crooked to allow any further progress. It is next to impossible to straighten a hole with a revolving core drill when once deflected; but should this occur with a percussion drill the remedy is very simple. On the principle that a suspended weight tends to point perpendicularly from its point of support the gravity percussion drill constantly seeks to retain or regain the vertical.

4. Where horizontal cleavages are present in the rock, a continuous core cannot usually be recovered by the revolving core drill; e.g., if a portion of the core becomes detached from its anchorage by a lateral cleft in the rock, it starts whirling with the core barrel. It is thus ground endwise under the water pressure against the part yet anchored, wearing away both itself and the part remaining fixed to the bottom. The materials thus ground off are washed down and out and lost. For this reason an accurate test of a soft coal vein, for example, cannot be made with a revolving core drill. Such veins are always full of cleavages and, the material being soft, the core is very fragile, easily jarred loose and broken off and readily ground up. True, a sample piece of the core may be recovered but, since the amount

worn off is wholly uncertain, the test for accurate thickness, or even of the general composition of the vein, cannot be made. At least the result would lack that certainty which would justify the expense of shafting. The cuttings and grindings, wholly or in part, may be washed away without being recovered at the surface, or may be so finely pulverized that, in a deep hole, they will float in the water unobserved. The same difficulties are met with in testing for granulous iron ore, lead, zinc and such minerals. These latter minerals usually occur in broken formations.

The Percussion Core Drill was expressly designed to overcome the above difficulties. In it the core barrel slips down over and protects the core as fast as made, and, since it does not revolve, it will not grind up the core, but will receive and retain it in the exact order of its production. Therefore with it a core can be recovered from a much softer and more broken formation than with any other form of core drill made.

5. A large hole cannot practically be made with the revolving core drill. The usual size of the core taken out is from 1 to 1¾ inches in diameter. Should a mud vein, a caving formation of clay or sand be found beyond the first rock, the smallness of the hole will not allow a casing to be inserted that further depth may be attained. On the other hand the large hole made with the percussion drill allows for such reductions.

6. A revolving core drill cannot be operated without abundance of water. A constant stream must be forced down the hollow drill rod. In rare cases this water may return to the surface for re-use; but even so it will be adulterated by washings from the sides of the well. Usually

it flows off through the fissures and is lost, carrying with it all the fine cuttings.

7. A revolving diamond core drill cannot be operated successfully without a skilled lapidary to re-cut and re-set the worn and loosened diamonds. The loss of a diamond is not only serious in itself, but it may cut the remaining points and destroy them before the loss is discovered.

8. The revolving core drill is operated by means of a pipe in threaded sections which run from the surface to the bottom. In operation this pipe must be jointed and unjointed every time the drill is put in or taken out. In deep wells this operation becomes very slow and tedious. On the other hand, the percussion drill, including our form of core drill, is operated by a stout manila cable fed off a geared reel as depth is attained. To lower the cable drill or draw it from a well of any depth is but the work of a few moments. The sand pumps are operated in the same way. It is true that, as a rule, the percussion drill must be withdrawn more frequently than the core drill—say at least every four or six feet—while a core drill may be run somewhat further than this distance without removal. But in mineral prospecting it is always an advantage, (no matter which kind of a drill is used), to run only short distances without examination of the findings.

9. What we have said above regarding the "revolving diamond core drill" is true also of the core drills employing chilled shot for the cutting. In the latter drills the cutting is done by means of a thick hollow tube, the end of which rides upon and presses the chilled shot into the opposing rock. If an open crevice be found, no matter how small, the operator must feed in enough shot to fill the crevice or

abandon the hole. Also if a seam be found that is filled with mud or joint clay, the shot is forced off into it in greater or smaller quantities. In revolution the hollow drill rod is likely to become locked to the standing core, break it loose and grind it up. As a matter of fact, no matter what sort of revolving core drill is used, a continuous section of core is not recoverable, except in hard rock, void of seams. But this does not apply to the percussion core drills described in Catalog No. 2-B.

These adverse contingencies, with others which they suggest, limit the legitimate and profitable use of revolving core drills to a specific and well defined field. We are constantly in receipt of letters which show that there is a wide misapprehension of their capabilities. Hence these remarks.

WHAT CANNOT BE DONE WITH KEYSTONE DRILLS

1. Nothing but a perpendicular hole can be made.
2. Except with our form of percussion core drill, no core can be recovered.

But aside from these two considerations the Keystone machines are universal prospectors.

It should be remembered that there is no formation through which our drills will not penetrate.

This is a broad, but fully warranted statement. It is based upon our twenty-five years successful, world-wide experience. It is certainly an advantage to the explorer to know beforehand that he will not find anything that will prevent his getting down. This is one very great advantage that our drills have over core drills of the ordinary type, or any of the lighter forms of cheap horse-power percussion drills.

Another very great advantage is that the cost per foot for sinking test wells with our machines is less than by any other method. As compared with core drilling, the cost per foot will perhaps not exceed one third. In other words, three separate six-inch holes can be made with our machine to the same depth, for about the cost of one two-inch test hole made with a revolving core drill. Where it is desired to determine the dip of a certain stratum it is necessary, no matter what kind of holes are drilled, to sink at several points.

It should be remembered that the large hole has another distinct advantage in that a greater quantity of material is recovered. The area of a core from a 2½" core hole is only about two square inches, while the cuttings from a six-inch hole made by one of our drills, represents an area of about 19 square inches, or nearly ten times as great. By our process only sufficient water is introduced to mix the cuttings, all of which water, under ordinary circumstances is again recovered by the vacuum sand pump, together with all the cuttings. The presence of a surplus of water is no hindrance. There is nothing about the tools easily broken and the operation requires but ordinary skill. Elsewhere we give adequate instructions for all ordinary work, showing how it can be done with accuracy and reliability.

Any of the machines described in this catalog can be used both for prospecting in rock formations and for placer testing by simply modifying the drilling tools to suit require ments. But it should be distinctly noted here that the drilling bits or "chisels" which are best adapted to solid rock drilling are not at all suited for use in alluvial prospecting. The former are too blunt and thick and if used, will have the effect of packing the cuttings off into the sides of the bore-hole beyond recovery. And, on the other hand, a drill bit which is suitable for placer prospecting is useless for drilling in hard rock. The blade of the drill bit used for alluvials being thin, it will inevitably make a three-cornered or "flat" hole in a hard formation. However, either style of drill bit can be used on the same drilling machine without changing any other item of the equipment in the least.

But the two processes are quite different, and to make this difference perfectly clear we shall describe both. We shall deal first with the Placer Testing Machines and their operation.

DIRECTIONS FOR THE OPERATION OF KEYSTONE MACHINES AND USE OF ACCOMPANYING APPLIANCES

By R. M. Downie
General Manager of Keystone Driller Company

MOVING AND SETTING UP

When the moves are short and the ground not too rough—the mast is not taken down. The Traction Machines move themselves almost anywhere without the assistance of animals, but the Non-traction usually require horses or oxen to move them, unless the moves are short. Short moves and steep inclines, unaccessible with horses, can be made by unreeling the drill cable (200 feet long), anchoring it to a tree or post and pulling by steam with the hoisting gear of the machine. All the machinery remains mounted at all times and is moved in its entirety. In swamps and quagmires it may be found necessary to place relays of cord-wood or planks.

The machine should be set so that the sills will be level and out of wind crosswise, but lengthwise they may be a little higher in front than in rear without detriment. By means of the wedges provided, level the front bolster until the crown pulley at top of derrick is plumb with the center-line of the machine. Reference to the illustrations on page 178 will show the appearance when set up and how the cable is rigged for use. The hinged derrick is raised to position by means of the long wooden braces. The derrick should lean a few inches forward at the top.

TO STRING THE DRIILLING TOOLS

For the purpose of ordinary placer testing 25 or 50 feet depth, the string of tools will consist only of three pieces; viz.: the "Rope Socket", Fig. 100; the "Stem", Fig. 102—immediately below it, and the "Bit", Fig. 116—below the stem.

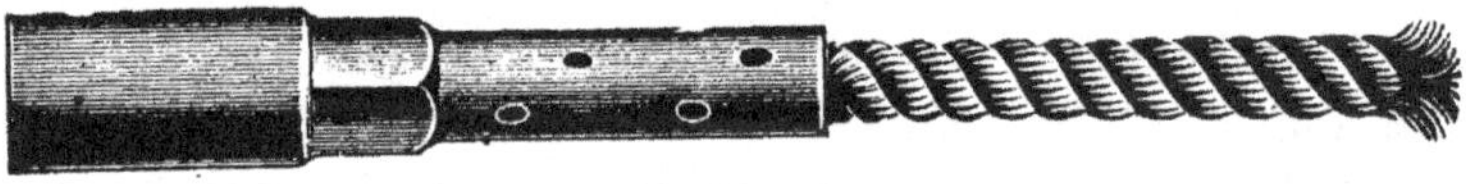

Standard Keystone Rope Socket—Fig. 100

The Jars (Fig. 101) need not be put on until the test well is 25 or 50 feet deep, and under favorable conditions it may not be needed at all. When used it is inserted between the Rope Socket (Fig. 100) and the Stem (Fig. 102). The joints on all of these tools are made of steel or the best grade of refined iron and are what are known as "taper joints." They are fully twice as strong as the old fashioned "straight joint." The collars are 4¼ inches in diameter. The "pin" (Fig. 108) and the "box" (Fig. 109) have squares for the tool wrenches. When not in use the threads are protected from dirt and injury by "thread protectors." They should be perfectly clean before putting together.

This tool is used to keep the bit from becoming fast in the well. Not needed until 30 or 50 feet is reached, and often not then.

The Rope Socket (Fig. 100) shows the drilling cable already in place—as it is when shipped.

To put the tools together, place the stem (Fig. 102) on the ground with the pin end of it near where the well is to be, and having drawn down a little slack cable, screw the rope socket on; also screw on the thin spudding-bit (Fig. 116) and set the joints up temporarily with the tool wrenches (Fig. 110) as tight as can be done by hand.

The Drill Stem is composed of a piece of round iron 4 inches thick and of a length and size to suit the machine. To the upper end of it is welded a male screw, or "Pin", Fig. 108. On the lower end is welded a female screw, or "Box", Fig. 109.

Jars—Fig. 101

This done, the engine is started and the entire tool is drawn up until the lower end of it is about 3 feet from the ground, and the brake applied to the cable reel to hold it suspended.

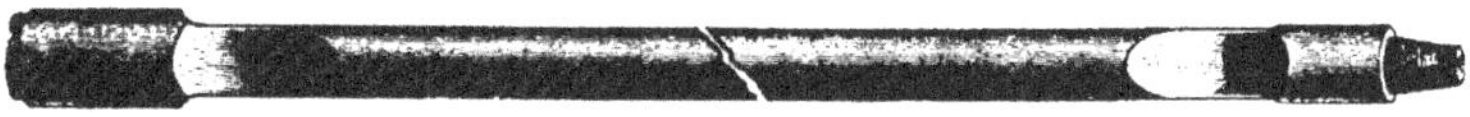

Drill Stem—Fig. 102

The Box and Pin are made in exact duplicates. There are many sizes of them manufactured, but those used on the machines shown in this catalog are known as "2 x 3 x 4—8 threads."

In subsequent operation, and when changing the bit for dressing, etc., there is a sleight in putting it on the stem. One man with the bit holder (Fig. 159), lifts the pin of the bit into the box of the suspended stem and holds it there while the attendant twists the stem until one or two threads

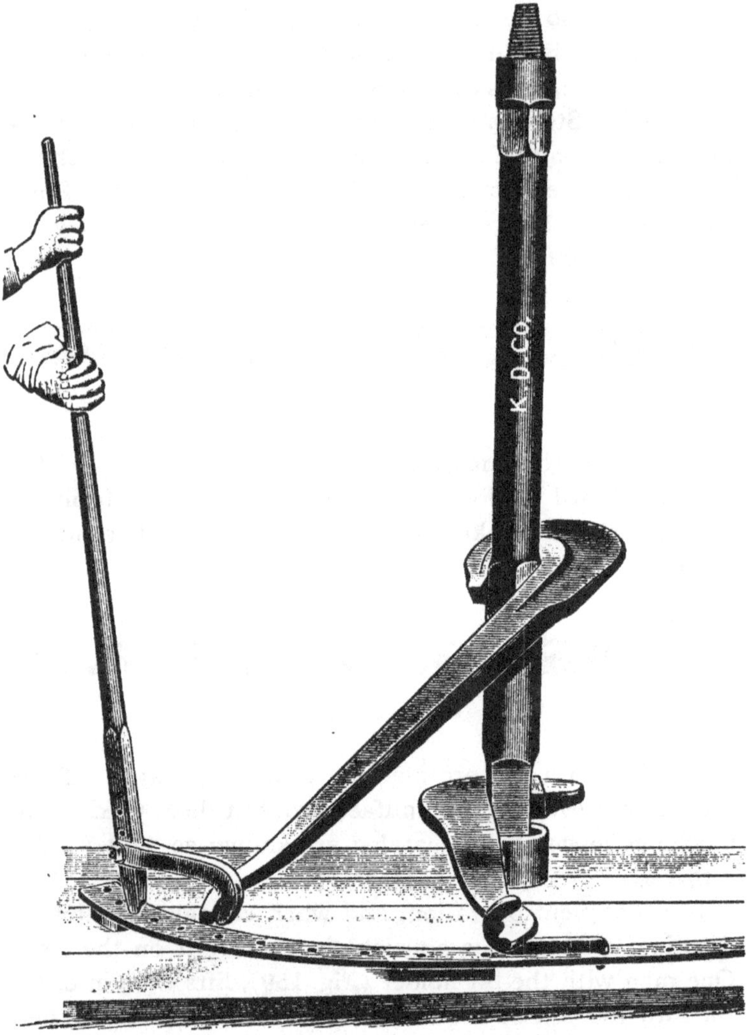

Fig. 110 shows method of using the wrenches with floor circle are engaged. Then the bit is turned by hand until the shoulders meet, the tool wrenches applied and the joints set up firmly. Before a joint passes below the surface it should be set up in a manner shown in the illustration below.

The wrenches should be put on the squares so as to pull toward the hook side as shown in the illustration. The upper one of the two in the illustration below is the "right-hand" wrench, and the one on the floor is the "left-hand." They are much stronger when thus used. The joints should be put together with all the force two persons can exert on the wrench bar. There is no danger of getting them too tight or of breaking the joint if made of the proper material. Never hammer the joint, either to tighten or loosen it, as this spoils the threads. About one-half of all fishing jobs arise from not screwing up the joints properly, and about two-thirds of the balance arise from using poor material in the joints, causing them to break in the well. In this respect, however, our customers have no trouble. Our joints are made with the utmost care and with the best material obtainable.

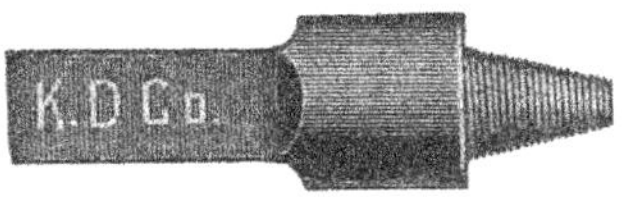

Pin Stub—Fig. 108

Box-Stub—Fig. 109

After a joint has been set up perfectly solid, a slight mark may be made across the joint with a sharp cold chisel, half of it on the pin collar, and half of it on the box. Each successive time this joint is screwed up, the mark on the box should go a little farther past the mark on the pin collar. In this manner it may always be known when the joint is

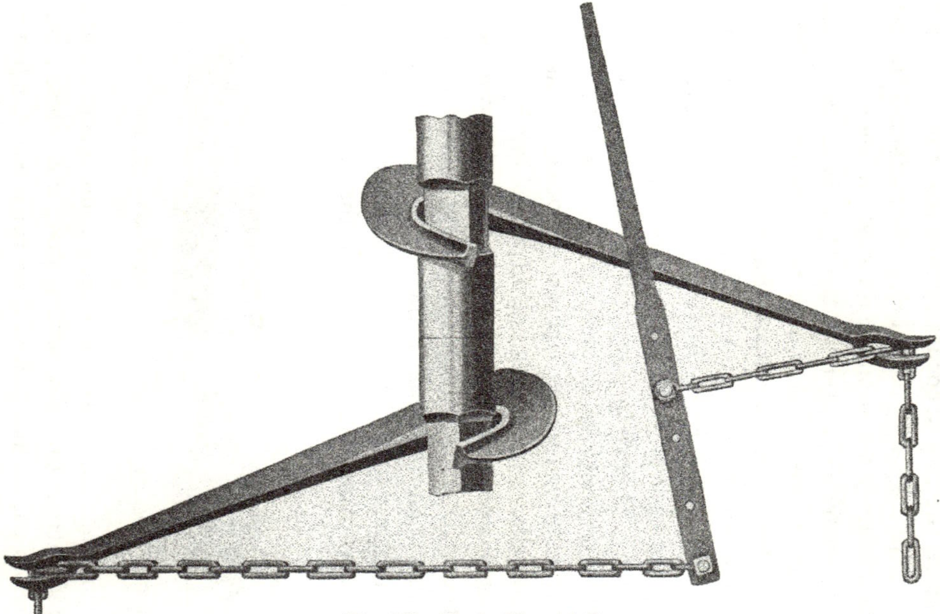

Fig. 118—Chain Wrench Bar

Note.—The tool wrenches with claw ends and chain wrench bar are now regularly furnished with placer testing and mineral prospecting outfits for moderate depths. If joints larger than 2¼″ x 3¼″ are used the double acting floor jack is preferable.

set up firmly. If at the next setting up it does not go far enough, you will know that there is dirt on the faces of the joint, or in the threads which should be removed.

When bedrock is reached, and if it be desired to drill some distance into it, or if a nest of heavy and hard boulders be encountered, the "spudding bit", Fig. 116, is taken off, and in its place is put on either a Mother Hubbard Bit, Fig. 104, or else a fluted rock bit like Fig. 105.

Fig. 104 is used only where the rock is full of wide seams and fissures, or where the strata may work out and cause the joint to become loose in the well. One mark on the box will be sufficient, each bit having one to match it. The perforated iron shown on the floor is known as the Floor Circle.

Fig. 110—Tool Wrenches like the above are used. The method of using the Wrenches is shown in the illustration on page 138.

When making short moves it is not necessary to take the drilling tools apart, the whole tool being carried upon brackets provided for it on the side of the machine.

Being nearly as wide at the top as at the cutting point, and the steel being very thick, the bit so nearly fills the hole that it cannot slip off sidewise into slanting openings to make a crooked hole.

Since there are only special localities where such a bit as Fig. 104 is necessary, it is not furnished with machines unless specified in the order. It will not drill as rapidly as Fig. 105, because the edges of the bit fit the hole more closely and cause more friction. With the equipment of any machine in this catalog it will be substituted for Fig. 105, without change in price, when desired by the purchaser.

For drilling in rock, Fig. 105, the Fluted Rock Bit, is generally used and is always furnished unless otherwise specified. By a reference to the Equipment List on a succeeding page, it will be seen that there are three bits furnished regularly with each placer testing outfit. Two of these are Fig. 116, the Thin Blade Placer Bit, and one of them is Fig. 105. The bits Fig. 116 are used in common for sinking the test tube, but if a heavy bed of boulders is found, or if it be desired to drill into a hard rock, Fig. 116 is taken off and Fig. 105 substituted.

It is a mistake to use Fig. 105, however, in strata of gravel and sand. This bit being designed originally for water and oil well drilling, where speed is the prime requisite, was made heavy and blunt. In a gravel or sand formation it will inevitably pack the material in its pathway and drive a considerable quantity of it off into the sides of the well, thus to a certain degree destroying the accuracy of the test. Fig. 116, is designed especially for placer prospecting, is made thin in the blade like a carpenter's chisel and will never pack the sand or gravel off into the sides of the bore hole.

During the spudding operation with which the well is begun the cross piece on the front of the walking beams, which carries the temper screw when in use, is unbolted and moved back a few inches. It may be entirely removed if desired. If left in place it will interfere with the stem and

KEYSTONE CUT DRIVE PIPE

In Placer Gold Testing it is desirable that the pipe, or test tube be of the most substantial quality; that it be cut in short lengths for ease in handling; and that the couplings and pipe be so threaded that the ends of the tubes will butt in the middle of the couplings so that there may be no danger of stripping the threads, in heavy driving.

For this purpose we use only the best grade of Extra Heavy Drive Pipe. The cutting and threading are done under careful supervision in our own factory. Drive Shoes, driving heads, etc., are forged and machined by us and threaded to accurately fit the pipe. Since it is of the utmost importance that all joints be perfectly matched and interchangeable we suggest that pipe to be used with our prospecting drills be purchased through us, particularly if it forms part of equipment to be shipped abroad or into distant and inaccessible places where replacement would be slow and difficult.

The cut herewith shows a test hole, including drill stem, drive clamps, drive head, couplings, a short length of pipe, drive shoe and placer bit.

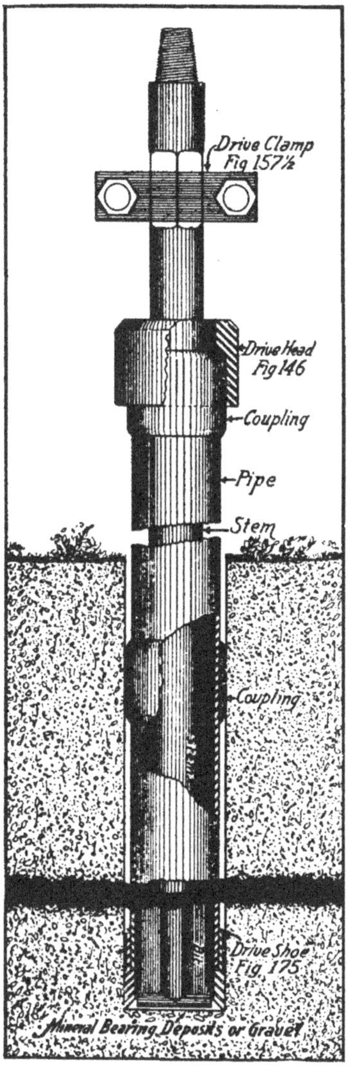

Fig. 104—Mother Hubbard Bit

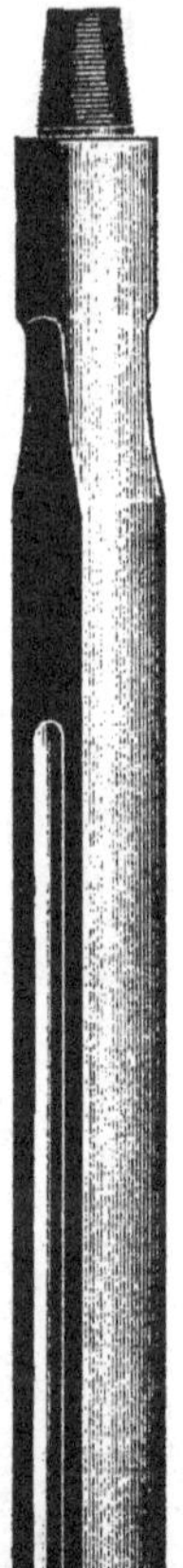

Fig. 105—Fluted Rock Bit

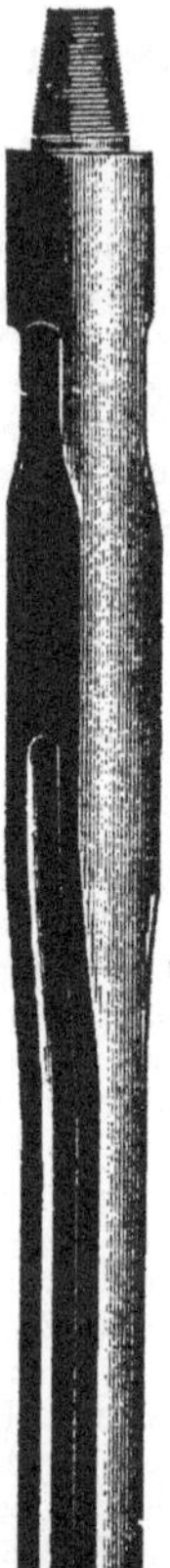

Fig. 116—Thin Blade Placer Bit

cable and be battered to pieces. On our larger machines will be found a special device for shifting this cross piece into and out of place.

To start the well, proceed as follows: Swing the drilling tools, now consisting of bit, stem and rope socket, in the derrick and set the reel brake in such a way that the point of the drill bit will not quite touch the ground when the walking beams are at their highest limit and the drilling tools are accordingly at the lowest limit of their stroke. The engine and the countershaft driven by it will now be out of gear with all the other machinery. While the engine is at rest throw the geared crank wheel into mesh and see that the latch holds it there securely. Now start the engine slowly, and after the weight of the drilling tool has taken up the slack cable, with the brake lever let the drilling tool down until it will barely touch the ground at the limit of its downward stroke. This will mark the spot where the well is to be made. Until the operator has learned to "balance" and steady the drilling tool from the surface, the attendant may temporarily take a position behind and on one of the cross-pieces of the derrick and steady the top of the drilling tool until a start has been made. Now start the engine, slowly at first, until the operator "gets the swing of the tool", and slacken the brake a little to let the cable run off far enough to allow the drill bit to touch the ground. A few blows will suffice to make an impression and start the hole. A little water should be supplied as needed, and as the hole deepens allow the cable, an inch or two at a time, to run off the reel. The operator, standing beside the well during this time, will steady the drilling tool, confining the blows to a central point and turning the tool a little at each blow so as to make

Fig. 160—Common Sand Pump

Fig. 161—Vacuum Sand Pump—Flat Bottom

Fig. 230—Improved Vacuum Sand Pump with Drop Bottom

a round hole. But in case the material at the surface is composed of loose sand or gravel, it may be found advisable to excavate with a shovel a hole about two or three feet deep and in this plant a 4 or 6-foot section of the drive pipe, with the drive shoe screwed firmly on the lower end and a sleeve coupling on the upper end. Drive first where possible, as it usually is. Sand and gravel may be tamped in about this section of drive pipe to hold it plumb, and in such case the drilling tool must be drawn up and entered over the top of the drive pipe. Proceed 2, 3, 4 or 5 feet, or until the cuttings become too thick to allow the drill to drop freely; then throw the crank wheel out of gear and the cable reel into gear and draw the drilling tool up until the bit is 2 or 3 feet above ground, and hold it there with the brake. When the Friction-geared Machines are used this can be done without stopping the engine, but when the Cog-geared Machines are used the engine must be stopped before throwing the cable reel into gear. Swing the bit out of the way, holding it with a loop or cord while the sand-pumping is done.

The common sand pump, Fig. 160, may be used to clear the hole of the cuttings. But if the soil be marsh or sand, having a tendency to cave in, the drive pipe should be driven down as far as the drill has penetrated.

The cuttings in the form of mud or mortar, enter the pump at the lower end and a valve prevents them from flowing out when the pump is raised. If the mud in the well be so thick that the sand pump will not readily sink through it to the bottom, raise the pump a foot or two and let it fall repeatedly until it is filled with the cuttings. It is then drawn up and emptied and the process repeated two or three times, or until the drill hole is entirely cleared of cuttings. A pail or two of water is then put into the drill hole, the drilling tool again lowered and the process repeated.

With each outfit there is furnished, in addition to the regular sand pump, Fig. 161, our Vacuum Sand Pump. The latter is made in two styles, with side dump, Fig. 161 and, with drop bottom, Fig. 230. These two styles of the vacuum sand pump are used for the same purpose and purchasers may take their choice at about the same price. The difference between them is that Fig. 161 must be turned bottom up in order to empty it, while Fig. 230 is unloaded by tripping the latch which supports the hinged bottom. Fig. 230 has the larger opening through the valve and will therefore admit larger gravel; and owing to the fact that the whole bottom drops out of the way, it will handle coarse, chunky and sticky material more easily. But, on account of its form, it must be used with more care than Fig. 161. When being lowered against a solid rock or boulder formation it should be let down gently and not dropped recklessly. The suckers of both pumps are packed with leather or rubber discs held in place by washers and lock nuts. This packing must eventually be replaced by the operator.

The Vacuum Sand Pump contains a valve or "sucker", which travels the whole length of the pump. When the pump is lowered into the well, this "sucker" goes to the bottom of the pump, being forced down by a heavy iron sucker-rod within it. When the sand reel is thrown in gear this sucker is drawn up rapidly and produces a vacuum in the lower part of the pump. The sand, clay, water and small stones are drawn into the pump by the inrush until it is full to the top.

When the sucker is lifted, the vacuum formed, and the contents of the well drawn in, the pump (per force of the vacuum formed) sinks and it will often bury itself two or three feet in the formation at the bottom of the well provided the material is not too solid.

The great advantage of the vacuum pump over the common one is at once manifest. It will pull into itself anything and everything small enough and loose enough to be taken. If gold or other minerals are present they are swept in with the current of mud and slush. The common sand pump will not do this, but will simply take out what will, by *gravity,* flow up through the valve. The operator must determine which pump is to be used at the surface—where the principal object usually is to simply clear the hole. But when a point is reached at which minerals are expected the vacuum sand pump should always be used.

The efficiency of the vacuum sand pump depends upon being lifted *quickly*. To this end the sand reel is operated by a powerful and rapid friction gear. The sand line should be securely attached, and in such a way that it will not cut where it is fastened to the sand pump. It is not uncommon that a lift of 2,000 pounds or more is required to start this pump if it has pulled itself 2 or 3 feet into the sand and gravel.

When testing for lead, zinc and iron ore, or in fact any mineral, this vacuum sand pump will fetch up everything in the hole. Its efficiency may be tested by throwing a handful of bird-shot into the well and sending the sand pump after it.

Where the well is to be started in a river bed or lake bottom, in a swamp where the water comes to the surface or in sandy soil which will not stand up at all without being cased, it will be necessary to drive the pipe from the very first. It is always advisable to do this.

THE DRIVE PIPE

The Placer Testing Machine, being generally used where there are boulders, sunken logs and débris, is equipped for using six-inch (inside diameter) drive pipe for hard driving. This drive pipe, unlike the ordinary Merchant Pipe, is made with straight threads, and the ends, being cut square off, meet in the center of the coupling. See Fig. 203.

For convenience in handling, this pipe is made in lengths of from 5 to 7 feet.

Since the amount required varies with different localities, this drive pipe is not included in the price of the machine, but is furnished by us as an extra in amounts and lengths to suit.

In Placer Testing this drive pipe must in all cases be driven to bedrock so as to insure that an exact area of test hole be carried clear to the bottom. Hence it is absolutely necessary that there be sufficient pipe at hand to reach the entire depth of the alluvial exploration. It cannot be driven into the solid rock, nor does it require to be further driven after bedrock is reached. But, while the depth of the alluvial deposit will determine the amount of pipe absolutely required, it is a common practice with our customers to order from two to three times this amount, so as to have a supply from which to replace injured sections. The joints, couplings and drive shoes are all made interchangeable and additional sections can be ordered from us at any time. After a test is completed the drive pipe is withdrawn; it may be used over and over again.

There are two weights of drive pipe—alike in quality, but differing in strength, weight and thickness. The lighter weight is about 5⁄16 inch thick, and is sufficiently heavy for all ordinary purposes. The "X Heavy" is about ½ inch thick, and is furnished for the most difficult work, where there are many boulders, or when great depths must be driven. The outside diameter of both the light and the "X Heavy" is the same, and the same sleeve couplings and drive shoes will fit both weights. Owing to its thicker walls, the actual inside diameter of the "X Heavy" grade is only about 5⅝ inches. Also the same size of drilling tools and bits are used with both weights—the bits being dressed by the operator to suit. Purchasers may therefore take any proportion desired of either weight.

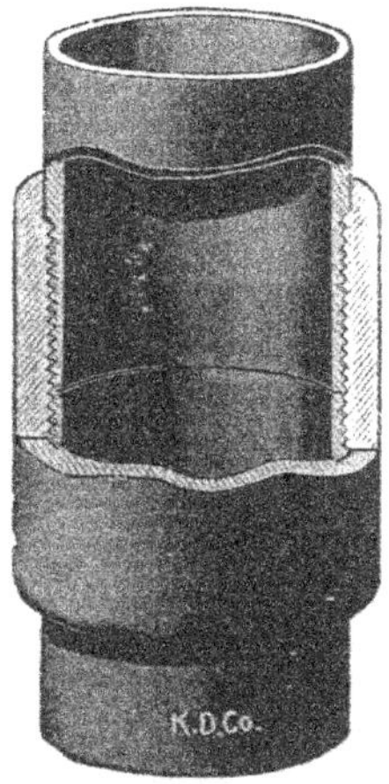

Drive Pipe Coupling

When the test is to be made from a boat in a lake or river bed, enough of the drive pipe is screwed together to reach

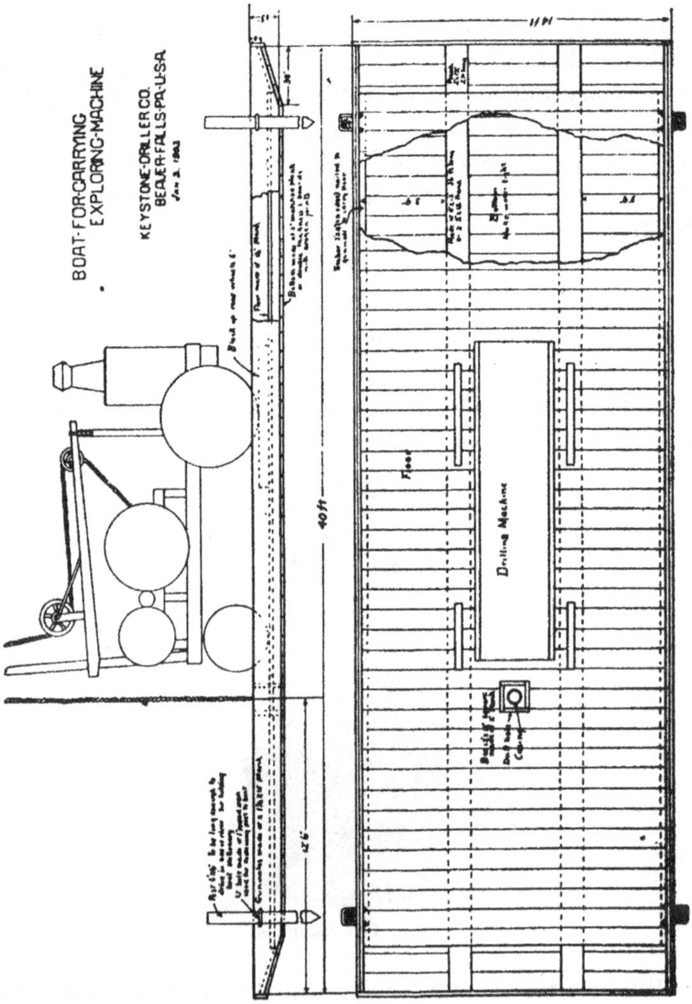
BOAT-FOR-CARRYING
EXPLORING-MACHINE
KEYSTONE-DRILLER CO.
BEAVER-FALLS-PA-U-S-A
Floor
Drilling Machine
40 ft

the bottom of the water and extend up 2 to 6 feet above the floor of the boat. When the test is made in ground which must be cased from the surface, one short section (say 7 feet long) is set up at the point where the well is to be made, with the drilling tool inside of it.

EXPLORING FROM A FLOAT OR FLAT BOAT

The machines described in the catalog being self-contained, are admirably adapted for use on flat boats for exploring lake beds, river channels or water covered marshes.

We do not cumber this catalog with detail plans for construction of the boat, but we will, without charge to purchasers of machines, furnish full plans and specifications. The lumber for boat construction can usually be had more cheaply at the place where the work is to be done—thus saving transportation charges. However, we will upon request quote prices upon the materials all cut to size, to be loaded upon same car with machine. The weight of the materials for a 40-foot boat 15 inches deep is about 8,000 pounds, and will with a machine and outfit of pipe make something over a minimum car load.

In use the boat is securely anchored with lines or spars over the spot to be prospected and through a conductor raised level with the top of the boat, the drive pipe is let down to the bottom. The depth of the water is immaterial except that the deeper the water the more pipe will be required. When the pipe has been set on the bottom the process is precisely similar to that described for prospecting on land.

When the test has been made the drive pipe is withdrawn by means of the Pipe Jars (included with regular Placer Testing Outfit) and the boat floated to next location.

The tests made in this way have been proved absolutely reliable and some very rich deposits have been found. The pipe pulling apparatus and the means of operating it make these machines peculiarly adaptable for this class of work.

To protect the lower end of the drive pipe from injury by being driven against boulders or slanting ledges, or being injured by the sharp corners of the drilling bit, and as well to insure that an exact area or section be excavated, a wrought steel Drive Shoe, Fig. 175, is used. It is usually made $7\frac{1}{2}$ inches in diameter at the cutting edge, and this represents the area excavated.

Fig 175—Wrought Steel Drive Shoe

It is slightly beveled inward and has a tempered edge. It is threaded to receive the 6-inch drive pipe, and has a shoulder against which the end of the pipe rests, as shown by the cut-away section in Fig. 175.

It is of prime importance that the drive shoe be firmly screwed on the pipe. It is made large enough to allow the pipe couplings to follow. In testing placer ground for gold, it is also of prime importance that this shoe have a perfect edge so as to cut an exact area, and that the pipe be driven so that at no time will the drill bit project any considerable distance below it. Otherwise, especially if the formation be of a caving nature, more than the area of the shoe will be excavated and an exactness of assay may not be obtained. For this reason the operator should have several of these drive shoes with his outfit, say three to five.

DRIVING THE PIPE

Whether the pipe is to be driven from a boat or started from the surface, as in a swamp, or whether, as in dry clay ground, a few feet has been first drilled, the drive pipe with the shoe in place is set as above described, plumbed and

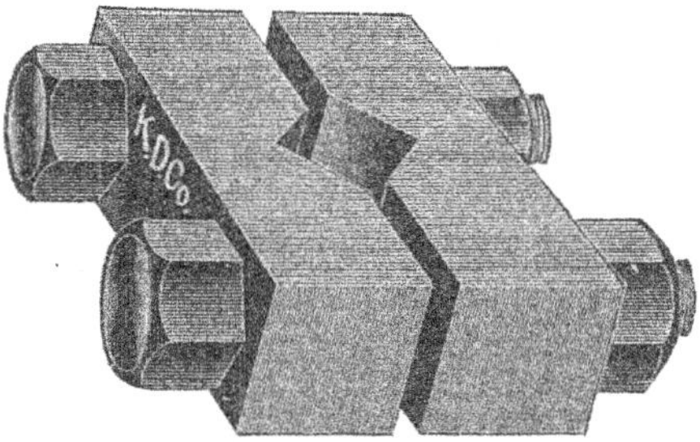

Fig. 157½—Driving Clamps

drilling is proceeded with. For a time the pipe may go down without driving, and were it not that the accurate test is desired the pipe could be let down quite a distance by drilling ahead of it. But if it does cave in so that the volume of material taken out with the vacuum sand pump is greater than the contents of the hole as represented by the cutting edge of the drive shoe, the accuracy of the test at that level will be spoiled. Therefore it is better to drive the pipe ahead wherever the material is soft. But it should also be remembered that it is possible to drive the pipe too

Driving Pipe

far ahead of the drill. If driven too far the materials may pack in the end of the test tube instead of rising into it. In such a case part of the material which should be taken out is driven off into the sides of the hole and not recovered, interfering with the accuracy of the assay.

In that part of the distance to be sunk in which no gold is found, it is immaterial whether the pipe is driven ahead of the bit or not, so it is gotten down. As a matter of fact it goes down more quickly and easily when the drill is kept well ahead, when the only object is to get the pipe down. Operators will therefore, have to use their discretion and be governed by circumstance. In any case the Driving Clamps, (Fig. 157½) are put on the stem as shown in the accompanying illustration. The illustration also shows the Driving Cap in place on the coupling.

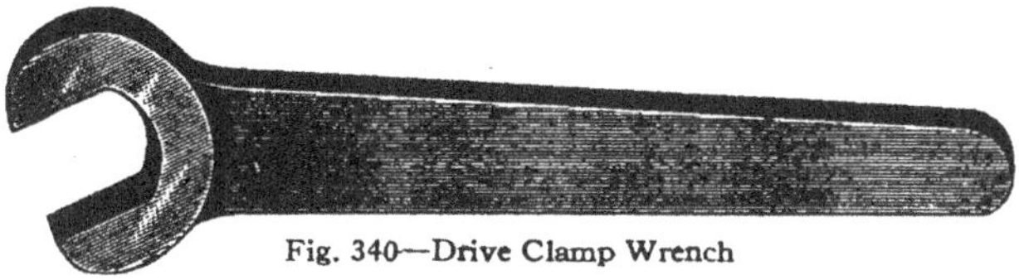

Fig. 340—Drive Clamp Wrench

The Driving Clamps (Fig. 157½) are, for common, put on the square at the upper end of the drill stem, the bolts being firmly tightened. But this square on the upper end of the stem being about 13 feet from the point of the bit, unless the hole is at least 13 feet deep (measuring from the top of the pipe), it may not be possible to use the driving clamps on the upper square of the drill stem. In such case attach them to the stem at the place prepared for them at mid-length of the stem until they can be used at the top.

In placer prospecting the Keystone Special Drive Head (Fig. 173) is used to best advantage. It is the simplest form of drive head obtainable, being made of a coupling and a nipple of extra heavy drive pipe. The cut entitled "Prospecting Crew in the Field", shows this drive head in place.

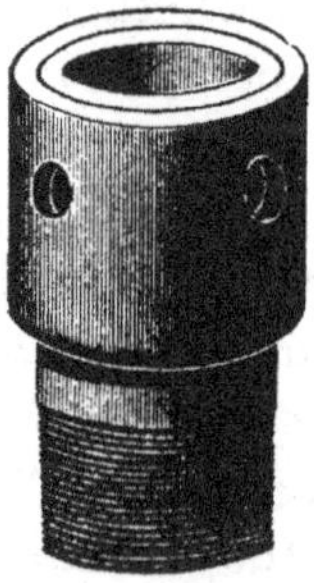

Keystone Special Drive Head

The pipe is driven by the weight of the drilling tool, by striking the driving cap with the driving clamps. The stroke of the drill (about 30 to 36 inches) will represent the length of the blow and the weight of the entire drilling tool (about 800 lbs.) will be the heft of the hammer. If the pipe drives hard place a matting of old rope, or any such substance, on top of the iron driving cap to cushion the blow. After driving 1, 2, 3 or 4 feet, according to circumstances, remove the driving clamps by taking out the bolts. Let the drill to the bottom and cut up the material in the pipe, adding water if need be.

Then use the sand pump as before directed. All material taken from the test tube should be carefully saved and panned.

A record of the length of each piece of pipe driven should be kept in order that it may be known exactly how deep the test tube is at any point of the proceeding, otherwise it may be difficult at times to determine whether the drill bit is working in the pipe or ahead of it.

The drilling bits should be dressed to the neat size of the bore of the pipe and the cutting edges should be kept as nearly as possible in the form in which they are when sent from the factory.

When putting the drive pipe together, clean the threads with a brush and lubricate them with a mixture of graphite (stove polish) and linseed oil. In no case allow any oil or grease to get inside the drive pipe, as it is liable to work down and be mixed with the materials cut up—preventing the possibility of panning out the gold. The accuracy of the test may be badly spoiled by allowing oil or grease to get into the hole.

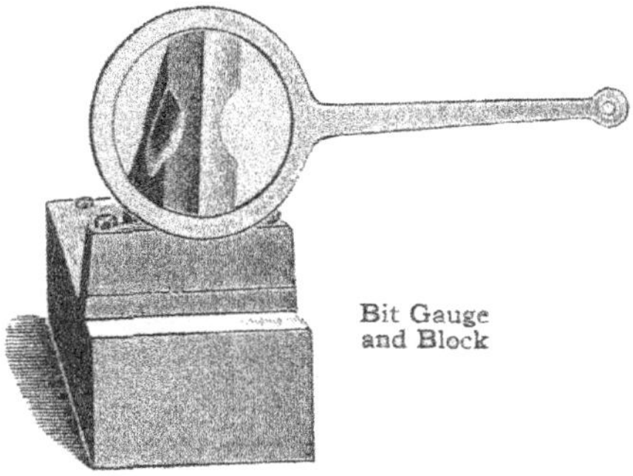

Bit Gauge and Block

DRESSING THE DRILLING BITS

The instructions given herewith are *general* and cover both rock drilling and placer testing.

The correct form for the bits will be learned from these cuts. Also the bits are correctly dressed when sent out, and their form should be particularly noted, and one may be kept for a model until the operator can duplicate it.

This cut (Fig. 154) illustrates a bit dressed for drilling a round hole in hard rock, and shows how the corners should fill out the gauge. The distance from **A** to **F** should be the same as from **C** to **D** or **B** to **E**.

To dress the bit attach one end of blast hose to the spout of the fan. The other end has on it a piece of two-inch pipe. Thrust the pipe firmly into the "tuyere iron." Within reach of the blast hose make a pit in the ground about the depth and size of a large wash basin, and place the tuyere iron in it. For heating the heavy bits this arrangement is better, lighter, cheaper and handier than an elevated forge.

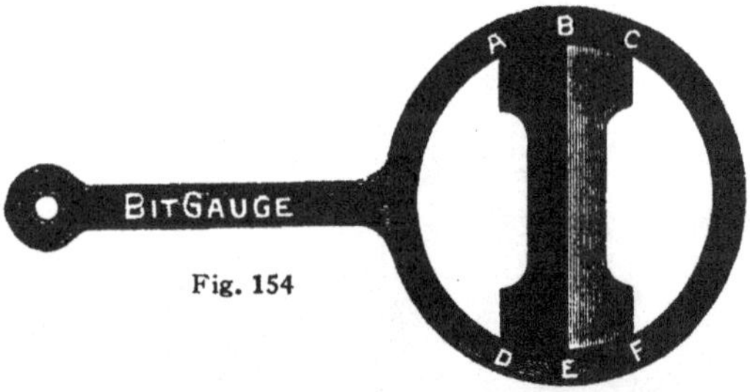

Fig. 154

Use soft coal for fire, if you can get it, but hard coal or even charcoal may be used for a bit dressing. Heat the bit to a cherry red for a distance of three or four inches back from the point, turning it occasionally in the fire to get both corners alike hot.

CAUTION

Drillers often make a mistake in not heating the bit far enough back. If a bit is heated at the point only, the hammering of it spreads the surface, but not the center.

By use of the "spectacles", Fig. 159, drag it upon the anvil block, and with the sledge spread it to a size somewhat larger than the bit gauge. Begin striking in the center and follow out to each corner. Turn the bit over and hammer the other side in the same way.

The "Spectacles" or Bit Holder

The next cut illustrates an improperly dressed bin for hard rock, but it is just right for placer testing work. It hard rock such a bit is likely to drill a three-cornered hole.

Fig. 158—Anvil Block

Turn bit on its edge with the lower corner projecting a little over the anvil billet, and make the corners A B C and D E F cutting edges. The corners of the bit do nearly all the cutting. The bits should conform to the circle of the gauge at the edges when the gauge is put on it at one-half

Tempering Bit

inch from the point. See Fig. 154. Keep the channels of the flute (see Figs. 103, 104 and 105) clear by using a cold chisel if in dressing, the steel laps back into the channel. Keep the bit straight across the face or cutting edge from B to E.

In smithing the Placer Bit, Fig. 116, the tool dresser should avoid making a blunt, cutting end like that used on Fig. 105 in hard rock drilling. This bit should be dressed with a sharper and thinner cutting edge, more like that shown in the illustration, Fig. 304.

To temper a drill bit, heat evenly to a very dark cherry red in the sunlight, and to a lighter cherry red in the shade or on a cloudy day or at night, but never heat steel until it throws off sparks. The sparks indicate that the carbon is being burned out of the steel, and it is thereby destroyed. See that the entire face of the bit is of even heat for about two inches back from the cutting edge, then set the bit plumb in not over one and a half or one and three-quarters inches of water. See Fig. 159.

With a stick, keep the water circulating for a minute until the point of the bit is cooled off. Tilt the point of the bit out of the water, and after rubbing the hammered part with a piece of stone or brick, to remove the scale, you will see a succession of colors creep gradually toward the point of the bit, as the heat runs down toward the end. The foremost one may be nearly white, the next "straw color" or orange, the next a deep yellowish "purple", then "blue", and finally black. The "white" is too hard for any kind of rock except soft slate, etc. The "straw" or orange is likewise too hard for hard rock, and will cause the edge and corners of the bit to crumble. The deep "purple" or "blue" is about right for the hardest rock, and is the temper used for mill picks, etc. The "black", of course is too soft, and a bit tempered to that color will soon lose its cutting edge and batter like a piece of iron. So soon as the purple or blue runs down to within three-fourths or one-half inch of the cutting edge, tilt the bit back into the water, stand plumb, and let it cool off. After the operator has tempered these drills a few times he will learn to gauge by sight the proper heat to produce the proper temper, and after this he will not need to take the bit out of the tempering tub to watch the colors.

The tool dresser should see that the threaded end of the bit is entirely cold before it is put on for drilling, otherwise the screw may shrink and loosen the joint after the tools are lowered into the well. Remember that,—

One-half of the art of drilling is in knowing how to dress the bit properly.

The reason for not tempering in deeper water is that the heavy body of steel will not quickly cool to the center. The outside shell will cool, contract and it may even crack open. The cracks thus formed become deeper at each heating and cooling, until pieces may spall off. The better the steel and the higher the carbon, the more likely is this to occur, but it can easily be avoided by tempering in shallow water as directed above, and only tempering the part which has been hammered. Always dress the bit out to the full gauge of the hole you are making.

The Clay Socket (Fig. 198) is used to take up soft materials intact. It is useless where boulders exist or in hard materials. But it frequently happens that overlying the bedrock where gold is found there is a plastic silt, clay or volcanic ash. In such places this tool is very satisfactory.

It works inside the 6-inch drive pipe, and is operated by being screwed on the drilling stem instead of the usual drilling bit.

PULLING THE PIPE

When a test well has been finished, the drive pipe used should (for economy) be drawn to be used over and over again till worn out. This is done by means of a pair of Keystone Pipe Pulling Jars, Fig. 500.

This tool is used for raising out of the ground the pipe which has been driven. It can be used in combination with the pulling ring and lifting jacks described on a following page, or it can be used alone. In use, the drill bit and stem are taken off and the Pipe Jar screwed on in place. The threaded knocking head (Fig. 501) is screwed into the coupling on the upper end of the pipe to be drawn. The machine is then set in motion and the tool drawn up so that the ram will strike against the lower side of the knocking head.

Our Special Pipe Clamp, Fig. 174, is then fitted on the pipe below the first collar and is used to keep the pipe from sinking back as it is raised by the Pipe Jar.

The Placer Testing Machines are friction geared so that the cable can be taken up as the pipe rises, without stopping. The weight of the tool serves as a hammer and the jarring will easily start a line of pipe which may be too firmly held to be drawn with the pulling ring and jacks alone. As the pipe rises the operator simply takes up the slack of the drill-cable until a full length of pipe is above ground. When the pipe is not too firmly held, this tool will raise it without the assistance of the jacks and do it very rapidly. The pin of the pipe pulling jars is threaded to fit the rope socket, and

the knocking head is threaded (unless otherwise ordered) to fit the coupling of six-inch drive pipe.

It will be noticed that the knocking head (Fig. 501) has a square opening. Into this the squared shank of the tool drops and one of the drilling tool wrenches, (Fig. 110) is then used to screw or unscrew the threaded knocking head from the drive pipe. The knocking heads are hardened against battering and the whole tool is durable and handy.

A special necessity for the tool is found when pipe has to be driven and pulled from a float or boat—when exploring river or lake beds, or testing under deep water for bridge piers, etc. In such cases the power required to pull the pipe by jacks is usually too great for the buoyancy of a boat or float, while with the pipe jars the pipe is raised by blows. Pipe that has been driven fifty feet into river silt and sand has been raised from a light flatboat without the use of jacks. It will be made to fit any desired size of pipe, and can be made so the same knocking head will fit two or more sizes of pipe, if so ordered.

PIPE PULLING RING

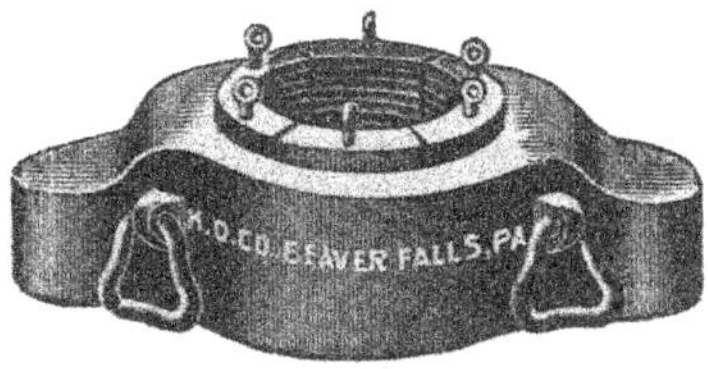

Fig. 190

There is also included with the Placer Testing equipment the Pipe Pulling Ring and Wedges (Fig. 190), and a pair of jack screws with which to operate it. The ring is made of

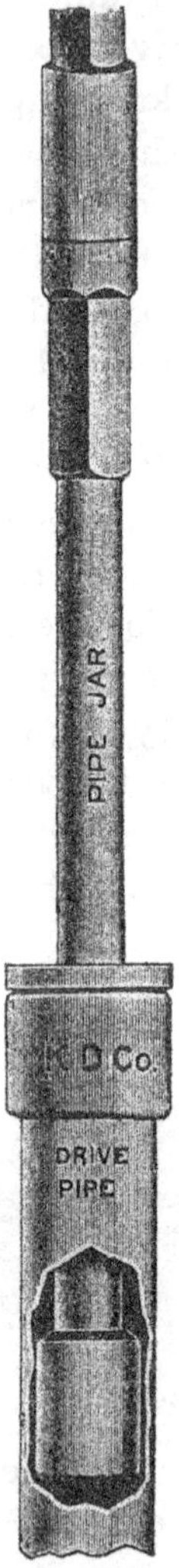

Keystone
Pipe Pulling
Jar

Fig. 500.

Fig. 501.
Knocking Head

cast steel with wrought steel, tempered, gripping wedges, which will catch the pipe at any point. It is used principally to start the pipe from its bed, if need be, after which the pipe pulling jars will draw the pipe in a fiftieth part of the time the screws could be operated.

The use of the various other appliances found in the Equipment List will be understood without explanation.

No. 3 Traction Friction Hoist

Machine at work testing for zinc near Galena, Ill

THE SCIENCE OF ZINC AND LEAD PROSPECTING WITH THE CHURN DRILL

By R. M. Downie
General Manager of Keystone Driller Company

Drilling a hole into the ground is not necessarily prospecting. Indeed, it is possible to drill a six-inch hole right through a ledge containing lead and zinc without at all detecting their presence. This was fully demonstrated some years ago in the Joplin District. Indeed, there was a time in that field when land owners would no longer lease their ground if it was to be prospected with a drill, for it had been fully proved by shafts afterward sunk, that much ground that had been condemned by improper drilling, actually contained paying quantities of mineral.

It has now been proved that the cheapest and best method of prospecting for these minerals is by means of the "Churn Drill", provided the proper kind of appliances are used. The only other methods are: First—By digging shafts; and, Second—By means of the "core" or "diamond" drills. The sinking of a shaft is a slow and expensive process and involves the disfigurement of the surface so as to permanently injure the ground for agricultural purposes. Theoretically, the "core drill" would be the ideal prospector; but, unfortunately its cost is too great, its operation too expensive, (three or four times that of the churn drill), and, finally, it is practically inoperative in the sort of broken and conglomerate formations in which zinc and lead are found in paying quantities. Prospectors are therefore practically confined to the use of the churn drill. When scientifically constructed, and intelligently handled, it serves the purpose admirably, doing the work cheaply, rapidly and reliably.

As we have noted at the outset it is possible with a churn drill to sink a hole through a vein of lead or zinc, and, through ignorance or lack of proper appliances, fail absolutely to detect its presence. It is possible also to drill through a mineralized stratum and obtain only a small or uncertain percentage of the minerals, thus making a deceptive and unreliable test, and, perhaps, appraising as next to worthless, ground that may be very rich. Unless all the mineral lying in the path of the drill and within the area excavated by it is brought to the surface, the test is next to valueless. If as little as ten per cent of the ore is lost, it may make all the difference between a profitable and an unprofitable prospect. Or, exasperating and tantalizing doubts may remain to tempt further expense or loss in shaft sinking.

Again, it is possible to drill a mineralized stratum of small or medium thickness in such a way as to make it appear two or three times as thick as it really is. Mistakes of this kind can be made by honest ignorance, and thousands of them have occurred in the Joplin Field, causing the loss of much money in sinking non-paying shafts and in running expensive drifts after mere streaks.

TWO PRIME REQUISITES IN A PROSPECTING DRILL

A Long Quick Stroke. A Vacuum Sludge Pump

QUICK STROKE—HOW OBTAINED

The first necessity in a prospecting drill is a long, quick stroke—the quicker the better. All of the percussion drills in use, including the Keystone, operate the drilling tool by gravity; that is, they raise the drilling tool on a cable by various sorts of crank, lever, trip or treadle mechanisms and let it fall of its own weight. This being true, the only thing that can be done to accelerate the down-stroke is to drop the tool with a perfectly slack cable, which the Keystone does in a manner peculiar to itself. This maximum speed of down-stroke being secured, it is equally important that the machine have a quick and powerful up-stroke, and in this respect the general speed of the stroke can be increased rather more indefinitely, depending upon the power and promptness of the operating mechanism. The Keystone is so constructed as to lift the tools in one-third of a revolution of the drive wheel. During the remaining two-thirds of a revolution the engine precedes the falling drilling tool. Thus, about 60 thirty-six inch strokes can be delivered per minute.

TENDENCY OF DRILLINGS TO SETTLE ON BOTTOM

Lead is about eleven times as heavy as water and pure zinc is about seven times as heavy as water. Compared with sand or clay, bulk for bulk, lead ore is about five times as heavy as wet sand, clay or cut-up rock, and zinc ore

("Jack") is about three to four times as heavy as the cuttings of the rock in which it is found. If, therefore, lead or zinc ores are broken, as is always more or less the case when cut up by a drill and mixed with sand, clay or cut-up rock and water, and allowed to settle, the lead and zinc will always be found at the very bottom of the vessel. This is exactly what happens in the bore hole when these minerals are struck with the drill unless prevented by the action of the drill itself.

TRITURATION OF DRILLINGS BY SLOW MOTION DRILLS

In the process of drilling, enough water must always be supplied to make a sort of batter or "sludge"; but generally it happens that water is found in the drill hole, above or in connection with the mineral largely in excess of what is actually needed, so that the cuttings are thoroughly saturated and washed, and if given any chance to do so, the mineral cuttings will immediately and during the process of drilling seek to settle upon the bottom of the drill hole as noted above. So heavy are they that, with a slow motion drill (30 to 40 strokes per minute), the cuttings will even settle between each two strokes of the drill and the descending drill-bit will, in that case, cut, recut and pulverize the chips so finely that most if not all are likely to be lost.

HOW TO PREVENT DRILLINGS FROM SETTLING ON THE BOTTOM

The stroke of the drill must therefore be long enough and quick enough to prevent this settling. Practice shows that the length of the stroke should be 36 to 40 inches and the drill should deliver 60 to 65 such strokes per minute.

Most of the "well drills" now in use in the lead and zinc field, however, have a stroke of only 12 to 15 inches, and even with this short stroke the drill strikes only 35 or 40 times a minute and the plan of the machines makes it utterly impossible to obtain the proper long, quick stroke.

KEYSTONE DRILL-TRAVEL 360 FEET PER MINUTE

The Keystone strikes 60 three-foot strokes a minute. This means a total drill-fall, per minute, of 180 feet; a total drill travel of 360 feet. It means that the drill is lifted in one-half the time it takes it to fall with a slack cable. (A full explanation of how this quick and powerful up-stroke is obtained can be found in our No. 1 Catalog.) On the other hand, some well drills being used, taken at their best speed, will not give a drill travel of more than one-half or three-fifths that of the Keystone. In this respect it stands alone, lifting the drill during about one-third of its crank motion and then allowing it to descend with a perfectly slack cable during the remaining two-thirds of its crank movement. This accounts for its quick action.

A SIMPLE EXPERIMENT

If our statements that a reliable test cannot be made with slow-stroke drills be questioned, try the following simple experiment: Take a weighed pound of lead or zinc to a well where one of these ordinary outfits is in use. After the hole is cleaned out, throw in the ore. After a distance of two or three feet has been drilled in rock of average hardness, let the well be sludged out. You will probably be surprised to find that you will not be able to recover one ounce out of the 16 you put in.

The reason for this is that the comparatively slow and short stroke of the drill allows the metal to settle between strokes to be cut and recut at each blow of the drill until reduced to a powder so fine that it will float imperceptibly in the sludge. And, if the pound of ore used in the experiment had been encountered in the path of the drill instead of being purposely put in, the presence of fifteen out of sixteen would never have been revealed.

ANOTHER SIMPLE TEST

Count out 100 grains of bird or duck-shot and cast them into such a prospect hole. After the drill has been operated for a short time, say ten minutes, see how many grains of shot can be recovered. If you recover a single grain that has not been cut and flattened out, or if you recover one in ten grains in any shape at all, you will be lucky.

Try the same experiment with a Keystone.

When the drill-bit leaves the bottom with its long, quick up-stroke, it creates a displacement something after the manner of a pump plunger. Its rapid withdrawal causes a down rush of sludge, which picks up and carries afloat all the cuttings, thus preventing them from being pounded up on the bottom.

AN INCIDENTAL ADVANTAGE OF THIS QUICK STROKE

Incidentally, this also adds greatly to the speed of drilling and is in a great measure responsible for the fame of the "Keystone Gait", for whether there be mineral or not, the bottom of the bore hole is kept clear so that the drill-bit will strike the clear rock at each blow. It therefore costs less per foot to make the test wells rapidly and reliably than to make them slowly and unreliably. As speed, however, is a secondary consideration in the drilling of test holes, we do not dwell longer upon this point.

THE SUCTION SLUDGE BUCKET

The Second Prime Requisite in Accurate Test Drilling

The second necessity in a reliable prospecting outfit is a means of thoroughly cleaning the bore hole after each run of the drill-bit. No argument need be brought in support of this statement. If any part of the mineral among the cuttings be left in the bore hole the test will be vitiated by so much. The two simple tests we suggested above, if applied, will convince any one that it is absolutely impossible to thoroughly clear a bore hole of lead or zinc cuttings by means of the common sludge buckets.

STILL ANOTHER TEST

If there still remains a doubt about this sweeping statement, try this: Take a six-inch post hole auger and bore a hole three feet deep in a clay formation that will hold water. Into this hole put a mixture of clay, sand and mineral cuttings and mix them with water to the consistency of drilling sludge. Now take a common sludge bucket and operating it by means of a rope as is done in drilling, see if you can recover with it all the mineral cuttings put in. One such test will fully convince you.

FALSE ASSAYS CAUSED BY IMPERFECTLY CLEANING OUT THE HOLE

To leave any part of mineral in the bore hole not only robs the ground of part of its value, but these unrecovered

Drilling for Zinc

cuttings will mix with those of the next bit run and will inevitably lead to the conclusion that the mineral bearing ledge is thicker than it really is.

All this difficulty is set aside by using the Keystone Vacuum Sludge Pump. The principle of its operation, as distinguished from the common sludge bucket is that a vacuum is formed under its plunger, which creates a sharp suction through its bottom valve. The downrush of sludge on the outside of the pump, and simultaneously on all sides of it, strikes on the bottom of the bore hole, licks up everything loose and carries it up into the pump. It will pick up lead shot from the smooth bottom of a bore hole, and this should be a conclusive test that it will take up lead and zinc ore.

The operation of this pump, it is true, requires good cordage, and a quick action reel; but, given these, it can be operated upon any sort of machine.

KEYSTONE VACUUM SAND OR SLUDGE PUMP

EXAMINATION OF THE SLUDGE

Third—Next in importance to having the right machine and sludge pump, in order to make a test that is accurate and reliable, it is necessary to save all the sludge taken from the mineral producing strata. The reason for this we shall now try to make clear.

Of course, under the sharp blows of the long-stroke drill, the cuttings will be much coarser than those made by the short, and therefore lighter, strokes of the common well-drill, and, while the quick action of the drill which we have just described will keep the drillings afloat and largely pre-

vent their pulverization, yet, under even the most favorable conditions, some part of the ore will be powdered in the process of drilling. The lighter the stroke and the slower the motion the more the drill-bit will operate like the pestle of a mortar. And here it should be remembered that nearly all lead and zinc ores are crystalline in structure and very easily powdered, and that it is easily possible to reduce these ores to a powder so fine that a half pound of it will float almost invisibly in a pail of clear water. The sludge taken from the bore hole being in the nature of very muddy water or thin batter, it is inevitable that a greater or less percentage of the ore will be held afloat in it, and the thicker the sludge, the longer it will require for the powdered minerals to settle. It is the common practice of driller men to discharge the thin and lighter portion of this sludge upon the ground, saving only the heavier settlings—under the impression perhaps that these coarser cuttings contain all the minerals. This is a gross mistake, and a practice which, for the general good of the mining business, cannot be abandoned too soon.

Perhaps the most convenient method of treating the sludge is to discharge it into a prepared sluice box, say 12 inches wide, 10 inches deep and 8 feet long, made of surfaced inch boards with a cross partition 6 inches high, about 4 feet from one end, and another partition about 4 inches high, 6½ feet from the same end. It will thus have three compartments. Into the larger of these the sludge is discharged from the vacuum sand pump. The heavier settlings will remain in this apartment, but the thinner sludge will overflow the 6-inch partition into the next compartment, in turn, after dropping their lighter cuttings, will overflow into the last and smallest. In the discharge end of the box, there may be a series of auger holes for slowly draining off the muddy water. The partitions may be set in loosely between strips so as to be easily removable for final cleaning up. After allowing reasonable time for settling the top water should be drawn off and more water (easily supplied from the well) should be added to carry away the earthy matter. For preliminary examination the coarse settlings collected in the larger bin may be scooped up and washed—returning all the wash water to the sluice box, and afterward the entire contents of the sluice box can be handjigged, panned or assayed.

(End of Part Two)

DRILLING COSTS IN POTASH PROSPECTING

By E. E. Free*

During the summers of 1912, 1913 and 1914 the Railroad Valley Company of Tonopah, Nevada, carried out an extensive prospecting campaign for potash deposits in various parts of the Great Basin and especially in Railroad Valley, Nye County, Nevada. In that vicinity the existence of a buried saline body was suspected, and the prospecting took the form of drill holes designed to reach and explore the supposed saline horizon. Five of these holes were located on the mud flat or *playa* left by the ancient lake which once occupied the valley, and these five holes were sufficiently alike in materials penetrated and in general conditions to be possible of discussion together. Although the general conditions surrounding the work were somewhat unusual, it is possible that the cost data of the accompanying table may have some general interest. The prospecting proved negative, so far as potash is concerned, and has been discontinued.

The work here reported was done between June 20 and November 10, 1913, but during this period there were two shutdowns aggregating 45 days, leaving a total of 99 days' actual work. Field operations were in charge of D. H. Walker, superintendent of the Railroad Valley Company.

Railroad Valley is southeast of the center of Nevada in a sparsely settled desert country without towns, railroads or other facilities. All supplies for the work were hauled by wagon or auto truck from Tonopah, 120 miles west, or from

*Chemical engineer, 1105 Madison Ave., Baltimore, Md.

Ely, 75 miles northeast. Drilling was by the rotary jetting method, using a rotary attached to a steam-driven Keystone portable rig No. 5. Water circulation was maintained by two steam pumps supplied by a separate boiler. The entire outfit was already in the possession of the company and cost approximately $7000, including drill rod, all tools and general equipment, also camp equipment consisting of four tents, boarding house, office, wagons, water tank, etc. No depreciation or interest on this equipment has been included in the costs given. The holes were 5½ inches in diameter and were not cased except for 20 to 50 ft. at the top.

The materials penetrated were essentially the same in all five holes. The upper portions were in smooth clays with little or no sand. The hardness was somewhat variable and a few cemented layers were encountered, but never of such induration as to create serious obstacles to drilling. In their lower portions holes No. 2 and No. 6 encountered beds of solid, crystalline gaylussite[1] alternating with beds of clay. The crystalline material was fairly hard and drilled like soft sandstone or shale. Similar beds were encountered but not entered by hole No. 4. The thicknesses of the beds penetrated by the individual holes are given in the table.

The analysis of costs given in the table requires a word of explanation. "Moving" includes the cost of transporting the rig and equipment to the site of the hole, setting up, and preparing for drilling. In the case of hole No. 2, this item covers the cost of moving the rig to the mud flat from a former location a few miles north. The expense of this first

[1]The mineral gaylussite is the hydrous double carbonate of sodium and calcium (Na_2CO_3 $CaCO_3$ $5H_2O$). It is a member of the saline series for which search was being made. For details concerning the occurrence see Free, "Min. and Sci. Press," Aug. 2, 1913.

COST DATA OF PROSPECT DRILLING

Hole number	2	3	4	5	6	Total	Average
Feet in clay	780	770	762	795	906	4013	803
Feet in gaylussite	63	none	none	none	84	147	...
Total depth, feet	843	770	762	795	990	4160	820
Days moving and setting up	18	13	9	10	13	63	12.6
Days drilling	6	10	5	5	10	36	7.2
Total time, days	24	23	14	15	23	99	19.8
Moving costs							
Labor	$440.60	$340.20	$239.70	$150.75	$292.85	$1464.10	$292.82
Team	633.40	103.90	106.00	16.00	76.92	936.22	187.24
Total	$1174.00	$444.10	$345.70	$166.75	$269.77	$2400.32	$480.06
Water supply costs							
Labor	$521.15	$561.02	$284.90	$133.90	$124.45	$1625.42	$325.08
Team	19.50	......	......	......	......	19.50	3.90
Total	$540.65	$561.02	$284.90	$133.90	$124.45	$1644.92	$328.98
Drilling costs							
Labor	$604.10	$747.30	$243.00	$340.70	$540.55	$2475.65	$495.13
Fuel	537.25	246.25	264.50	214.35	431.75	1694.10	338.82
Casing	20.70	24.84	27.60	27.60	26.22	126.96	25.39
Total	$1162.05	$1018.39	$535.10	$582.65	$998.52	$4296.71	$859.34
General expense							
Supplies and repairs	$146.81	$140.70	$85.65	$91.76	$140.70	$605.62	$121.12
Camp maintenance	163.25	156.45	95.23	102.04	156.45	673.42	134.68
Boarding-house deficit	95.07	91.11	55.45	59.42	91.11	392.16	78.43
Communications	325.84	312.24	190.06	203.64	312.24	1344.02	268.80
Superintendence	312.00	299.00	182.00	195.00	299.00	1287.00	257.40
Chemical work	313.40	300.33	182.81	195.88	300.33	1292.75	258.55
Liability insurance	21.33	20.45	12.44	13.33	20.45	88.00	17.60
Total	$1377.70	$1320.28	$803.64	$861.07	$1320.28	$5682.97	$1136.59
Total costs	$4254.40	$3343.79	$1969.34	$1744.37	$2713.02	$14,024.92	$2804.98
Cost per foot, total	$5.047	$4.343	$2.584	$2.194	$2.740		$3.371
Cost per foot, drilling only	1.378	1.323	.702	.733	1.009		1.033

moving was much increased by continuous bad weather. Water supply for drilling and for the boilers was a matter of extreme difficulty. No water was available at or near the drill sites and preliminary 100-ft. holes had proven dry. When a preliminary 250-ft. hole (sunk with hauled water) at the site of hole No. 2 also proved dry, it was decided to bring water by ditch from an artesian well seven miles north. Water supplied by this ditch was used for all subsequent holes. In the table the item of water-supply for hole No. 2 includes the cost of the main ditch and of all preliminary work. For the other holes only the cost of ditch maintenance and of necessary extensions is included. The cost for hole No. 3 is high because the extension of the ditch to it had to be carried through a sand ridge nearly a half-mile wide and several feet high.

The item of drilling includes the actual cost of drilling only. The usual crew was: One driller, one helper, one fireman and one sampler. . Occasionally two helpers were needed. Usually there were two drilling shifts of ten hours each, the firemen working three 8-hour shifts and looking to the pumps during the 4-hour shutdown. For a part of hole No. 2 three 8-hour drilling shifts were used. In addition, one or two men were employed on repairs and general mechanical work and one to three teamsters were hauling drinking water and supplies and keeping the ditch in order. The wage scale was $6.50 to $5.00 for drillers and $5.00 and $4.50 for helpers, fireman and teamsters, all on the basis of an 8-hour day with pay for overtime at the same rate. Lodging in tents was furnished free. The boarding house was run by the company and a charge of $1.00 per day was made for board. The deficit on the boarding house appears under general expense in the table. Teams were hired at $7.00 per day for team and driver and $5.00 per day for extra teams, not found. Fuel was wood, no other being available at reasonable cost. It consisted largely of scrub pine and juniper, and was cut and delivered by contract at from $10.50 to $12.00 per cord.

General expense includes all items which extended over the whole work and cannot be assigned with accuracy to the different holes. These items have been totaled and the total divided between the holes on the basis of the working time for each hole. The item for communications includes the maintenance of the company's automobiles and the hire of other automobiles when necessary. The employer's liability insurance is that required by the law of Nevada. The item for chemical work covers the cost of an accurate chemical control of the drilling, this being necessitated by the object of the work. It includes the salary of a chemist at $7.00 per day and the expense of the field laboratory. The laboratory equipment was borrowed from the general laboratory of the company and is not charged for. It would have cost about $300 to duplicate it.

SUCCESSFUL SALTING OF ALLUVIALS

By C. S. Haley

In attempting a disquisition upon this subject, one is minded to take pattern after the famous essay upon snakes in Ireland, which, as will be remembered, began with the pertinent statement, "There are no snakes in Ireland." By the same token, then, there is no such thing as successful alluvial salting, provided: that the examining engineer is honest; that he takes the care, either through his own personal knowledge, or through some absolutely dependable recommendation, that all persons coming in direct contact with the handling of his samples are either honest or directly under the eye of some one who is; and that he use ordinary care, vigilance, and common sense in checking up the results of his sampling.

PRELIMINARY EXAMINATION OF ALLUVIALS

For an instance, I will take the easiest conditions for alluvial sampling; preliminary examination of an ancient river bed with a view to hydraulic installation. If the deposit is salted, it may lead to the expense of a more careful and, in inaccessible localities, more costly final examination, the result of which might mean a great loss of prestige. In a case of this sort, the examining engineer is apt to make the trip alone, or with one assistant at most; as in all probability his natural inclination, made pessimistic by force of circumstances and experience, is to believe that he is going on a wild goose chase, and should be accomplished with as little loss of time and money as possible.

Arriving at the neighborhood of the property, it is worse than useless for him to attempt to conceal the nature of his business. In these our United States, so neighborly is the interest felt in all the operations of our brother, the chances are that the proprietor of the hotel in which he rests his stage-worn bones on the first night of his journey, has been telephoned in advance of the nature of his errand, and in the course of his evening's wait he will have all sorts of information gratuitously offered him. If, however, our engineer be not too young he will not shun this information thus thrust upon him, but may derive a great deal of keen amusement therefrom. For in even the most isolated mountain communities the knockers thrive as well as the boosters, and he is very like to hear a composite description of the property which he proposes to examine which will be very like to a cubist picture; viewed from one angle, a radiant vision of a glorious dawn; from another, a dark-brown taste on a foggy morning. From such an extreme double characterization, by noting the points overemphasized, he can often determine a course of action, subject to modification on the ground, which will perhaps aid him to the saving of much time. On the other hand, he may decide it best to let it all pass with a "warming of the head", as our Latin-American friends very aptly put it, and forget all about it.

Arrived at his destination, and well aware that the object of his trip is a matter of discussion to the entire country-side, it may be well that he does not look upon the men whom he hires to attend to the rougher portion of the work as being like unto Caesar's wife. It may be necessary to run an open cut into the bank, and down along the rim of the channel; or, in the case of a working face, to strip a cut from top to bedrock. In all of this work, there is the easiest chance in the world to salt samples, provided that the engineer is careless and easy going. On the other hand, if he appears to be very insouciant, the chances are that by keeping a watchful eye beneath such an appearance, he can detect much more than if he preserves the air of extreme vigilance. Most salting in such places is very clumsily done. For instance, if you give a man a sample to pan down for you (intending, of course, to pay no attention to the result), you may observe him smoking with his pipe tipped sideways over the pan, or spitting tobacco juice nonchalantly into the pool in which he is panning. But this kind of thing should never be noticed openly. Simply give the man enough rope, and he will hang himself, and you will avoid a great deal of unpleasantness, and possibly the necessity of doing, or receiving, physical harm. One should always remember that the last laugh comes with the report. And, no matter who is handling the manual work of the sampling, always take check samples yourself, from a face exposed by yourself, and pan them yourself, with nobody else within five feet of you.

AN UNSUCCESSFUL ATTEMPT

So much for this type of sampling, which is very simple, and does not necessitate, in the case of an experienced engineer, any weighing of samples. In connection with this work, I am minded of a tale they still tell in Yreka of two gentlemen of fortune who attempted to dispose of a worthless placer claim to a Chinaman! After very carefully disposing of about five hundred dollars worth of wash gold about the exposed face of the workings with a shotgun, they offered the property to Wun Lung, who had recently disposed of his laundry and pi-gow house, for five thousand dollars cash. Wun Lung very cheerfully consented to enter into negotiations, but requested a day to work the property, to determine its worth. This was, of course, readily accorded him. At the end of the day, the prospective purchaser expressed himself as very well satisfied, but wished to wait two more days before paying over the five thousand. With the lure of easy money in their eyes, our gentlemen of fortune readily assented. On the time appointed, however, the Celestial failed to materialize; in fact, he could not be found in the town. The owners of the property, stricken with a sudden panic, went down to their diggings, and found their worst fears realized. The carefully scraped banks showed them whither had departed their $500 worth of dust, and they returned to town to walk in the virtue of the chastened.

This kind of thing is, of course, clumsy; so clumsy that it often defeats itself; as in a case mentioned in *The Mining Magazine* some years ago, where a young man went out to sample a gravel property, and, in spite of the fact that the promoters managed to slip at least ten cents worth of gold into every pan that he took, he reported the property as valueless simply for the reason that he panned all the gold out and even had no black sand left at the end of his panning!

KEYSTONE AND EMPIRE DRILLING

In the case, however, of Keystone and Empire drilling, much more artistic work can be done. For this reason, the panner on the drill, or at least one of them, should always be a man absolutely trustworthy and personally known to the examining engineer. In case of any irregularity on the other shift of the drill, it may always be checked by the work of the absolutely trustworthy shift. One very ingenious method of salting a Keystone that has come under my notice has been the plastering of the drill rope with mud containing plenty of No. 3 colors. In this manner the *de trop* gold was carefully and evenly distributed throughout the hole. A watchful drill-runner, who felt the fine flakes of mud dropping round him as they dried, spoiled this little plan very effectually, however.

A drill fireman with the tobacco habit spoiled his game once by spitting too consistently into the box as he lowered or raised the sand pump. The clumsiest attempt that I know of in this connection was the deliberate spilling of gold on top of the ground after the machine had been set up and the careful salting of panning tubs and boxes. The simpleton did not even take the trouble to use the same type of gold that was ordinarily found in the bar!

DRILL SAMPLING

In drill sampling, however, there is this big advantage over open-cut sampling; that you have a continual check on the weight of your gold by the panners' log. In open-cut sampling, where possibly samples are only weighed once in two or three days, unless the samples are very carefully guarded, it might be very easy to augment the sample; and, as the colors have only been scanned once, and then as a lump sum, a careful job might raise the value of the property several cents on the yard. But in either case, until all samples are weighed up and checked, the utmost care should be taken in guarding them. Usually, in the case of a property extensive enough to justify a Keystone examination, the amount of money involved is big enough to induce men of a certain type to take extraordinary chances.

Of course, in the long run, the most effective way of salting alluvial, as well as other property, is to buy the examining engineer; but this, too, presents its difficulties. If the report is to be worth anything in the eyes of practical mining men, it must be made by an engineer of standing. And, in most cases, that standing is the result of long years of honest work and toil-won experience. Therefore, the man who would barter it, even if he would barter that incomparably more priceless thing, his self-respect, would probably demand so high a price that the promoter's profit would be more than wiped out. For of necessity, to a man who has lived an honest existence as far as the maturity of his life, the prospect of not being able to hold his head high for the balance of his career, is hard to assess to him in financial terms.

In conclusion, one of the most effectual methods of salting a gold mine that ever came under my experience, was not done with gold, but with frogs! The promoter of the property, which was once a famous producer in "the days of old", etc., was making a strenuous effort to sell a worked-out mine. His prospective buyers were Frenchmen, and the whole incident came under my personal observation. The price was moderate, the gold was fairly evenly distributed, and there was a fair chance to sell. However, instead of making the mistake of trying to prove the impossible, my friend left the subject of available and workable ground strictly alone, and expatiated on the value and beauty of the place as a summer camp. Then he led his buyers down to a deep and spacious pool of stagnant water in the bottom of an abandoned, worked-out diggings. It was a sort of a frog concentrate, and the rest of the mine, from a frog standpoint, was not to be judged by it. At the opportune moment, nets were produced, and before his clients left that summer afternoon, they were begging him to accept a check to bind the bargain.

(SPECIMEN)

FIELD LOG

....Property

Line No............ **Hole No.**...........

Date Commenced........................192...... **Date Finished**........................192......

Depth in Feet	Estimated Weight of Gold SIZE			FORMATION	CORE		REMARKS
	1	2	3		Before	After	
							In Charge
							Driller
							Panner

Total depth..........feet **Bedrock**..........**feet** Water level..........**feet**

ABBREVIATIONS

T.	**Tailings**	S.	Sand	St.	Sticky
F.	**Fine**	G.	Gravel	Md.	Medium
C.	**Coarse**	Cl.	Clay	M.	Much
V.	**Very**	Sm.	Some	L.	Loose

RECORD OF FORMATION

USED IN CONJUNCTION WITH LOG BOOK

No. of Hole.................... Date,..192

....Contract

....Tract

DEPTH Feet	NO. OF COLORS Size 1	Size 2	Size 3	FORMATION	REMARKS
....	....	....	....		
....	....	....	....		
....	....	....	....		
....	....	....	....		
5					
....	....	....	....		
....	....	....	....		
....	....	....	....		
....	....	....	....		LOCATION
10					
....	....	....	....		
....	....	....	....		
....	....	....	....		
....	....	....	....		
15					
....	....	....	....		
....	....	....	....		.... In Charge
....	....	....	....		
....	....	....	....		.... Prospector
20					

Total Depth.........Feet. Bedrock.........Feet. Water Level.........Feet.

PRINTED IN U.S.A
THE EDDY PRESS CORPORATION
PITTSBURGH, PA.

Spotted something that isn't quite right?

Let us know and we'll fix it for future reprints. And as a thank you for your diligence, we'll send you a free Sothis book of your choice.

publisher@sothis.press

This page intentionally left blank

Made in the USA
Las Vegas, NV
06 May 2026

46469932R00121